# Manufacturing and Surface Engineering

Special Issue Editor

**Alicia Esther Ares**

MDPI • Basel • Beijing • Wuhan • Barcelona • Belgrade

**MDPI**

*Special Issue Editor*
Alicia Esther Ares
National University of Misiones
Argentina

*Editorial Office*
MDPI
St. Alban-Anlage 66
Basel, Switzerland

This edition is a reprint of the Special Issue published online in the open access journal *Coatings* (ISSN 2079-6412) from 2017–2018 (available at: http://www.mdpi.com/journal/coatings/special_issues/manuf_surf_eng).

For citation purposes, cite each article independently as indicated on the article page online and as indicated below:

Lastname, F.M.; Lastname, F.M. Article title. *Journal Name* **Year**, *Article number*, page range.

**First Editon 2018**

**ISBN 978-3-03842-979-1 (Pbk)**
**ISBN 978-3-03842-980-7 (PDF)**

# Table of Contents

# About the Special Issue Editor

**Alicia Esther Ares** has been Headline Professor of Materials Science at the Chemical Engineering Department, School of Sciences (FCEQyN), National University of Misiones (UNaM), Posadas, Misiones, Argentina, since December 2013. She has also been an Independent Researcher at the National Scientific and Technical Research Council (CONICET), Argentina, since January 2015. Previously a research associate at CONICET (2008–2014) and Associate Professor at UNaM (2007-2013), she was also an Assistant Professor at UNaM (1989–2007). She graduated from the University of Misiones in 1992 and completed a PhD degree in Materials Science at the Institute of Technology "Jorge Sabato", UNSAM-CNEA, Buenos Aires, Argentina. Later she held postdoctoral positions at the following institutions: Faculdade de Engenharía Mecânica, Departamento de Engenharía de Materiais. Universidade Estadual de Campinas. Campinas. São Paulo. Brasil (2001 and 2005–2006), Department of Materials Science and Engineering, University of Florida, Gainesville, Florida, United States (2002–2003) and the Faculty of Sciences, National University of Misiones, Posadas-Misiones, Argentina (2003–2004).

She has 29 years of teaching experience both at the undergraduate and the graduate level.

Her research interests lie in the areas of:

—Synthesis and characterization of nanostructured coatings, membranes and templates of aluminum and zinc oxides.

—Fabrication and characterization of nanostructured titanium and iron oxide coatings for water treatment systems based on advanced oxidative and reductive processes.

—Natural products as corrosion inhibitors of metallic materials.

—Solidification thermal parameters, mechanical properties and corrosion resistance of different alloys and composite materials.

—Solidification structures and properties of alloys for hard tissue replacement.

—Metallic material selection for the management of biofuels.

Her articles are published in well-established international and Argentinian journals: https://www.scopus.com/authid/detail.uri?authorId=7004713687.

*Article*

# Thermal-Sprayed Coatings on Bushing and Sleeve-Pipe Surfaces in Continuous Galvanizing Sinking Roller Production Line Applications

**Guangwei Zhang [1,2], Deyuan Li [1], Ning Zhang [1], Nannan Zhang [1,\*]and Sihua Duan [2]**

[1]   Materials Science and Engineering, Shenyang University of Technology, Shenyang 110870, Liaoning, China; lichaod@dhidcw.com (G.Z.); dmy1962@sut.edu.cn (D.L.); zhangning201610034@sut.edu.cn (N.Z.)

[2]   Dalian Huarui Heavy Industrial Special Spare Parts Co., Ltd., Dalian 116052, Liaoning, China; thjj@dhidcw.com

\*    Correspondence: zhangnn@sut.edu.cn; Tel.: +86-24-2549-6812

Received: 29 June 2017; Accepted: 28 July 2017; Published: 2 August 2017

**Abstract:** This paper describes thermal spray techniques for making hard coatings on bushing and sleeve component surfaces. Specifically, plasma-arc welding was used to produce 5-mm thick Co-Cr alloy welding overlays on the bushing, while a high-velocity oxy-fuel spraying technique and laser re-melting technique were used to produce thinner coatings of Co-Cr-Ni+WC of about 1 mm thickness on the sleeve-pipe counterparts. The surface-treated components were then submerged in liquid zinc to study the corrosive behaviour of the surface coating and substrate. Both the scanning electron microscope and energy dispersive spectrometer analyses were used to study the microstructure and phase composition of both coatings and substrates prior to and after corrosion experiments. The results show that the microstructure of the bushing consists of γ-cobalt solid solution as well as the eutectic structure of γ-cobalt and carbides, which have good corrosive resistance against molten zinc. Meanwhile, the microstructure of the sleeve pipe consists of a Co-Cr solid solution with various forms of carbides, which displays the combined properties of toughness with good corrosive resistance to molten zinc.

**Keywords:** sink roller; bushing; sleeve pipe; thermal spray; high-velocity oxy-fuel (HVOF) spray; welding overlay

---

## 1. Introduction

In the continuous galvanised zinc (Zn)-coating production lines in typical manufacturing factories, the rollers consist of sink rollers and tension rollers. Sink rollers are submerged in molten Zn baths and are subject to high-temperature corrosion in Zn solutions and continuous wear of the steel strip. The working environments of the rollers are extremely harsh [1–4]. Bushings and sleeve pipes are the main components on the sink-roller assembly, which are used as supports and rotation pairs for the sink rollers. Similarly, bushings and sleeve pipes are also subjected to high temperature, corrosion and abrasion. As a result, they generally have a very low in-service life, with its current working lifetime being approximately 15 days [5,6].

Sink rollers are normally made of 316L of stainless steel. In the past, the lifetime of sink rollers was usually about a week (or seven days) [7]. Recently, development has shown that with high-velocity oxy-fuel (HVOF)-sprayed carbon-deficient WC-12Co coatings, the lifetime of sink rollers had been extended to beyond 20 days. Nowadays, the challenge is how to extend the lifetime of the bushings and sleeve pipes, so that their service lives are compatible with that of the sink rollers. Bushings and sleeve pipes are also made of 316L of stainless steel. Studies have shown that the damage to these components was usually due to: (1) severe corrosion caused by molten Zn in the bath and (2) formation of Zn dregs that introduce severe wear problems on the interacting surfaces. The combined corrosion and wear eventually leads to rotation failure or fracture of the components [8,9].

To improve the service life of bushings and sleeve pipes as well as to reduce economic losses, many methods had been chosen to improve their wear resistance and anti-corrosion, such as selecting specialty materials, using a surface hardening process or other methods. The choice of materials usually includes high nickel (Ni)-chromium (Cr) stainless steel, Ni-based alloys or cobalt (Co)-based alloys for creation, but it obviously leads to higher manufacturing costs [10–13]. The method of material surface infiltration and centrifugal casting for the preparation of the bushings and sleeve pipes with a composite structure is a way to reduce the production cost and maintain excellent performance, but the manufacturing process is also very complex.

Surface modification and thermal spraying are effective methods for fabricating composite structures for bushings and sleeve pipes. These techniques had been effectively applied on the surfaces of strengthened sink rollers, which is the industrial standard today [14]. To date, however, little work has been done on the preparation of the surface hardening of bushings and sleeve-pipe surface techniques.

Therefore, this paper reports the recent works of surface modifications of Co-based alloy bushings and sleeves pipes using a combined welding overlay and the HVOF thermal spray technique. It also includes the surface properties of the coatings associated with their performance in a molten Zn work environment.

## 2. Experimental Procedure

### 2.1. Substrate Fabrication of Bushings and Sleeve Pipes

Using the Chinese industrial specifications, the bushings and sleeve pipes are made in-house at the Dalian Huarui Heavy Industrial Special Spare Parts Co., Ltd., Dalian, China, using 316L-type stainless steels. This is shown in Figure 1, which are the typical dimensions in the sink-roller assemblies of continuous galvanised Zn production lines in steel mills. Figure 1a shows that the bushing is a cylindrical-type of size 200 mm/100 mm × 100 mm, where the interacting surface is at its inner surface of the cylinder. Figure 1b shows that the sleeve pipe is also a cylindrical-type structure, with dimensions of 95 mm/70 mm × 105 mm, where the interacting surface is at its outer surface of the cylinder.

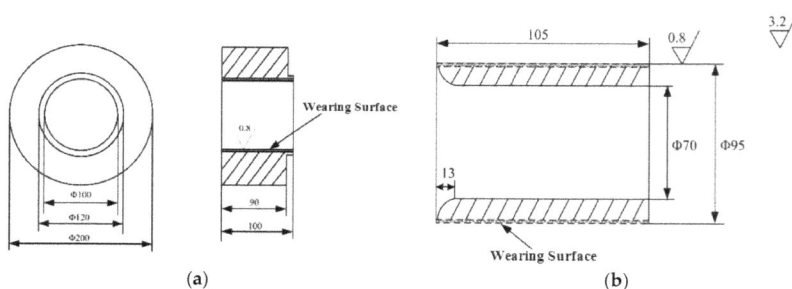

**Figure 1.** Schematic illustration of the typical dimensions of a sink-roller assembly at the continuous zinc galvanising production lines in steel mills: (**a**) bushing and (**b**) sleeve pipe.

### 2.2. Coating Feedstock Materials

Varied materials are used for the bushing and sleeve-pipe surfaces, as shown in Table 1. The feedstock composition for the bushing is a Co-Cr-based welding alloy consisting of 61.25%Co and 30.19%Cr as well as other elements including 4.08% tungsten (W), 1%Ni, 1.48% silicon (Si), 0.5% manganese (Mn) and 1.5% carbon (C). The feedstock materials for the sleeve-pipe surface coating is a mixture of the Co-Cr-Ni alloy and WC powder, with a ratio of nearly 4:1. Molybdenum (Mo) was added to improve the corrosion resistance and Ni could improve the adherence of coatings. The composition is shown in Table 1.

**Table 1.** Compositions of the powder for both bushing and sleeve pipe.

| No. | Feedstock Material | Chemical Elements (wt %) | | | | | | | |
|-----|--------------------|------|------|------|-------|------|-------|------|------|
|     |                    | C    | Si   | Mn   | Cr    | Mo   | W     | Ni   | Co   |
| 1   | Bushing Coating    | 1.50 | 1.48 | 0.50 | 30.19 | –    | 4.08  | 1.00 | Bal. |
| 2   | Sleeve-Pipe Coating| 2.45 | 0.55 | 0.89 | 29.50 | 0.45 | 17.50 | 8.07 | Bal. |

*2.3. Coating Fabrication*

We aimed to achieve a 5 mm-thick coating for the bushing surface. With such a thick coating goal and its inner surface nature or non-line-of-sight (NLOS) surface, neither plasma spray nor HVOF spray systems can accomplish the task. Therefore, the plasma-arc-welding technique was used to prepare the coating. For the sleeve-pipe surface coating, we aimed to produce approximately a 1 to 2 mm-thick coating, with its line-of-sight (LOS) surface geometry nature. A HVOF spray technique and laser-melting technique were used to produce the coating.

For the bushing surface coating, a welding overlay technique was used to prepare the surface coating for the bushing. The welding torch model is BX-ZD-400A, made by Benxi Mechanical & Electrical Technology Co., Ltd., Shanghai, China. The plasma welding parameters are shown in Table 2.

**Table 2.** Plasma welding operation parameters.

| Current (*A*) | Voltage (*V*) | Gas Flow Rate (L/min) | Cylinder Rotation Speed (r/min) |
|---------------|---------------|-----------------------|---------------------------------|
| 400–450       | 35–45         | 15–25                 | 60                              |

For the sleeve surface coating, JP-8000 (Praxair, Danbury, CT, USA) HVOF spraying equipment was used to prepare the surface coating for the sleeve-pipe surface. Before thermal spraying, sample surfaces were sand-blasted using alumina-based sand with a mesh size of 30, followed by de-greasing using acetone. The spray parameters are listed in Table 3.

**Table 3.** High-velocity oxy-fuel (HVOF) spraying operation parameters.

| Fuel Flow Rate (L/h) | Oxygen Flow Rate (m³/h) | Powder Feeding Rate (g/min) | Cylinder Rotation Speed (r/min) | Gun Travelling Speed (mm/min) |
|----------------------|-------------------------|-----------------------------|---------------------------------|-------------------------------|
| 21–23                | 55–60                   | 65–70                       | 60                              | 1000                          |

Since the sink roller is subjected to extensive loading (especially shear torque) during operation, the inner laminal structure and porosities of sprayed coating lead to poorer adhesion between the coating and substrate. Therefore, the laser-melting technique (LDF6000-100, Laserline, Mülheim-Kärlich, Germany) was used to re-melt the surfaces of the HVOF spraying coating, so that metallurgical coating bonding can be achieved and the splat boundaries between each layer are eliminated. The re-melting parameters by lasers are listed in Table 4.

**Table 4.** Re-melting operation parameters by lasers.

| Power (kW) | Ar Flow Rate (L/min) | Focal Length (mm) | Cylinder Rotation Speed (r/min) | Gun Travelling Speed (mm/min) |
|------------|----------------------|-------------------|---------------------------------|-------------------------------|
| 2.5        | 20                   | 400               | 60                              | 500                           |

## 2.4. Molten Zinc Corrosion Experiments

Regarding specimen preparation, using electrical discharge machines, the specimens were prepared by cutting the coated bushing piece into the dimensions of 15 mm × 15 mm × 10 mm (Figure 2a) and the coated sleeve-pipe specimen into dimensions of 15 mm × 15 mm × 12.5 mm (Figure 2b).

For the corrosion experiments, ceramic crucibles with aninner size of 92 mm × 118 mm (modelA5; and volume = 625 mL, Jintuo Experimental Equipment Co., Ltd., Changsha, China) were used. The samples were taken out at 72 h, 120 h and 168 h.

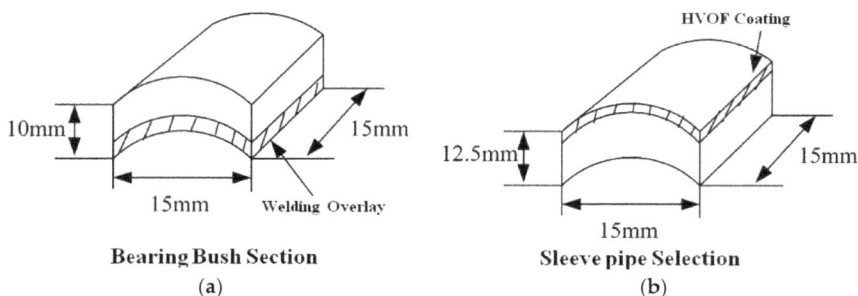

**Figure 2.** Schematic illustration of the specimen dimensions: (**a**) bushing with Co-Cr alloy coating and (**b**) sleeve pipe with Co-Cr-Ni+WC coating.

## 2.5. Coating Analysis

The coated and after-corrosion samples were cut into small cross-sections for metallography using epoxies for sample mounting. The mounted samples were then ground using a grinding machine (GPX200, Leco Corporation, St. Joseph, MI, USA), with the grinding started at a mesh size of 100, which was reduced down to a mesh size of 800. This was followed by polishing using a 1μm diamond slurry. An optical microscope (IA44+GX51RBB, Olympus, Tokyo, Japan) was used to analyse the coating thickness and porosity. The thickness of the coating was measured from the top of coating to the interface, with ten areas obtained to calculate an average length. The Vickers microhardness tester (HVS-1000, Shidai, Beijing, China) was used to measure coating hardness.

A scanning electron microscope (SEM; S3400, Hitachi, Tokyo, Japan) was used to analyse the cross-section samples to study the detailed microstructure of the coating and associated Zn-corrosive behaviour. An energy dispersive spectrometer (EDS; S3400, Hitachi, Tokyo, Japan) was used to analyse the overall coating composition and their localised phase identifications. In addition, X-ray diffraction (XRD; type 7000, Shimadzu, Kyoto, Japan) was used to study the phases of the coating prior and post-corrosion testing.

## 3. Results and Discussion

### 3.1. Coating Morphology, Phase Composition and Hardness

Regarding the bushing surface coating, the XRD spectrum is shown in Figure 3. Since no Cr and W phases were detected in the XRD spectrum, this probably indicates that Cr and W atoms are in the solid solution of the FCC Co structure, while the Cr-carbide phase is the hard phase dispersed in the Co-Cr matrix phase. The deposited Co-Cr alloy has a microhardness of approximately 530–580HV0.3. A Co-Cr (70:30 atomic ratio) would exhibit a microhardness of approximately 340–350HV0.3. This slight increase in hardness is probably due to the presence of the Cr–carbide phase-dispersion strengthening effect.

Figure 4 is an image of the welding overlay coating produced on the bushing surface. The two-phase microstructure is clearly shown, including a grey matrix phase and a dark bi-phase.

To verify the chemical structure of the two distinct phases, EDS was focused on both the grey matrix Area a and the dark dispersion Area b with their spectra shown in Figure 5a,b, respectively. The quantitative analysis of the spectrum for these two areas is listed in Table 5. The hard-facing layer of the sink-roll shaft contains Co, Cr, W, C, Si and iron (Fe) elements, among others. The composition is the same as the powder in Table 1, but with the Fe element, which could not be found in the powder. This may be produced by a parent metal in the process of welding. The Co content is higher in the primary phase a, while the C is very low. This can determine that it is a γ solid solution phase (γ-Co), which forms some solid solution with a certain amount of alloy elements, such as Cr and W. In the primary phase at the grain boundary eutectic b, the C content is higher due to being concentrated with Cr and W, which shows that the phase is $Cr_7C_3$-type carbides with a γ-Co eutectic structure.

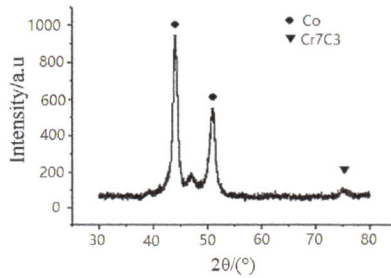

**Figure 3.** XRD spectrum of bushing surface coating.

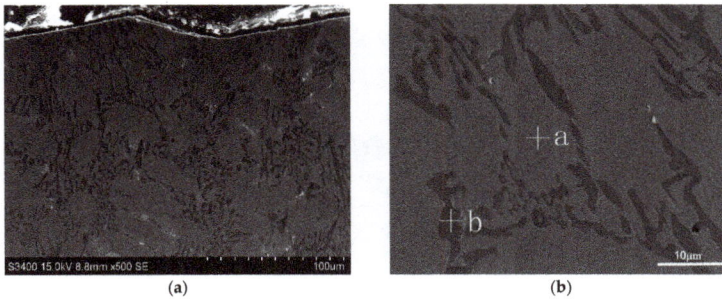

**Figure 4.** Morphology of the deposited Co-Cr alloy coating on the bushing surface: (**a**) cross-section of the Co-Cr alloy coating; (**b**) position of EDS.

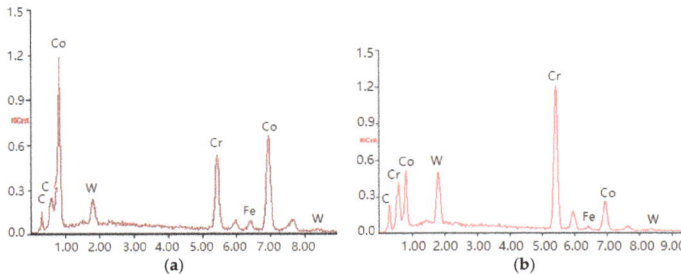

**Figure 5.** Presence of atomic elements by EDS in (**a**) Area a, and (**b**) Area b.

**Table 5.** Compositions of different areas in the produced bushing coatings by energy dispersive spectrometer (EDS).

| No. | Positions Analysed | Chemical Elements (wt %) | | | | | |
|-----|--------------------|------|------|------|------|------|------|
| | | Co | Cr | W | Fe | Si | C |
| 1 | General Area Average | 55 | 30 | 4.0 | 4.2 | 1.4 | 1.7 |
| 2 | Grey Matrix Phase Area a | 63 | 22 | 2.8 | 4.6 | 1.9 | 0.9 |
| 3 | Dark Phase Area b | 12 | 63 | 10 | 1.9 | 0.1 | 4.0 |

Regarding the sleeve-pipe surface coating, the XRD spectrum is shown in Figure 6 and the XRD spectra is quite complex. It indicated that the XRD spectra of the coating shows diffraction patterns of the following: (a) major FCC-Co phase with peaks of (111), (200) and (220) at 2θ angle of ~44°, ~51° and ~72°, respectively; (b) hexagonal W carbide (WC phase) peaks of (001), (100), and (101) at 2θ angles of ~31.5°, 35.6° and 48.3°, respectively; (c) η-carbide phase $Co_3W_3C$ at 2θ angles of ~42°; (d) a minor Cr-carbide $Cr_3C_2$ phase identified at 2θ angles of ~65°.

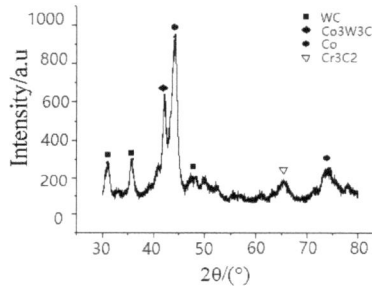

**Figure 6.** XRD spectrum of the sleeve-pipe surface coating.

The measured Vickers microhardness is approximately 900–1200HV0.3, indicating that the deposited sleeve surface is a complex Co-Cr-Ni alloy that has a co-existence of multi-phasic carbide phases.

Figure 7 is an image of the post-laser treatment of the HVOF-sprayed coating produced on the sleeve-pipe surface. It is clearly a multi-phasic microstructure, which includes a grey facet large particulate phase (Area a), a grey contiguous smaller precipitate phase (Area b), a large irregularly shaped crystal phase (Area c) and a dark matrix phase (Area d). The EDS spectrum for all areas is listed in Table 6.

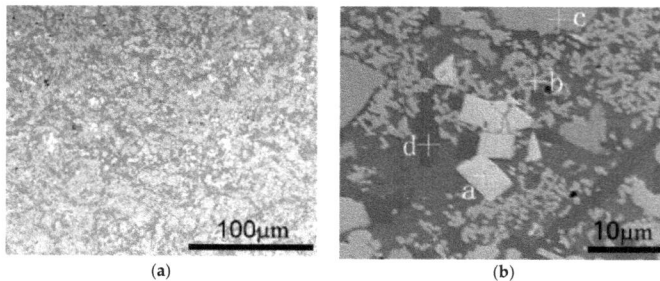

(a)　　　　　　　　　　　　(b)

**Figure 7.** Post-laser treatment of the HVOF-sprayed coating on the sleeve-pipe surface: (**a**) cross-section of treated coating; (**b**) position of EDS.

**Table 6.** Compositions of different areas in the produced sleeve-pipe surface coatings by EDS.

| No. | Positions Analysed | Chemical Elements (wt %) | | | | |
|---|---|---|---|---|---|---|
| | | Co | Cr | W | Ni | C |
| 1 | General Area Average | 28.0 | 15.0 | 44.0 | 8.0 | 3.7 |
| 2 | Grey Facet Large Particulates Area a | – | – | 90.0 | – | 9.3 |
| 3 | Grey Contiguous Smaller Precipitates Area b | 21.0 | 3.0 | 68.0 | 2.3 | 4.6 |
| 4 | Large Irregularly Shaped Crystals Area c | 17.0 | 10.0 | 58.0 | 7.7 | 5.5 |
| 5 | Dark Matrix Phase Area d | 32 | 44.0 | 12.0 | – | 2.9 |

The above results show that the coating composition and microstructure are relatively complicated. The alloy system is a Co-Cr-Ni-W series, which contains a higher proportion of Cr and Co solid solutions as well as a certain proportion of W. There are three types of precipitated phase distribution. The zone in Area a is a white WC phase, which should belong to the mix with the WC powder. The Area b zone is a grey compound carbide phase including Cr, Co, Ni and W, which should belong to the $Co_3W_3C$ phase. Furthermore, this compound contains some W elements in the γ-Co solid solution. The Area c zone has some small circular particles, which are distributed like "eggs". Its composition is mainly C and W, meaning that it should belong to the WC or $W_2C$ compounds.

### 3.2. Corrosion of Bushing Surface Coatings

The coating has an original thickness of 5 mm. Their after-corrosion thicknesses are 4.475 mm for 72 h, 4.15 mm for 120 h and 3.65 mm for 168 h, which leads to a coating corrosion rate of 0.008 mm/h. The after-corrosion samples at the sampled times for the bushing coating for 72 h, 120 h and 168 h are shown in Figure 8a–c, respectively. Figure 8a shows two regions after the 72 h corrosion, including: (a) the top region with the molten Zn adhered onto the coating interface and (b) the bottom region with the coating. During the Zn corrosion, the coating morphology does not change, a new phase is not formed, while corrosion channels or cracks are not formed in the coating. In addition, the fact that there is no observed corrosion rate difference between the light-coloured alloy phase and the dark-coloured particulate phase (or eutectic phase) is likely to be the primary reason there is a uniform corrosion mechanism without the formation of cracks or interconnected porosity channels for the welding-overlaid bushing surface coating in the corrosion experiments.

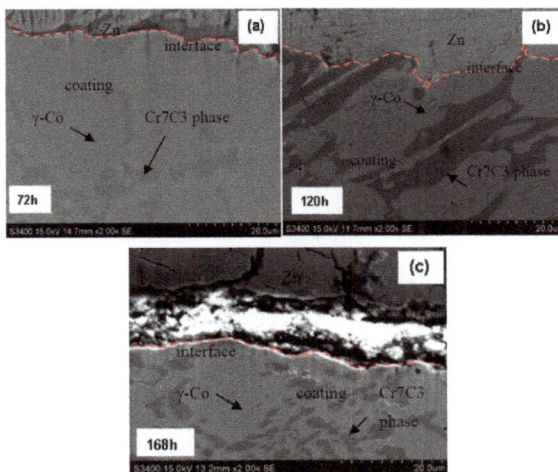

**Figure 8.** Microstructure of the bushing coating after corrosion tests at: (a) 72 h; (b) 120 h; (c) 168 h.

Figure 8b shows the SEM image of the sample after 120 h of corrosion. There is some evidence that the matrix phase revealed a slightly higher corrosion rate than the dark-carbide particulate phase due to the presence of carbide particles protruding at the coating/molten Zn interface. However, the preferred corrosive resistance of the carbide particles did not result in a dramatic change in coating corrosion rate or the formation of cracks or corrosion channels in the coating. This is probably because the volume fraction of the carbide phase is too small to cause a dramatic corrosion rate difference in the material. This preferred corrosion difference is also evident in the 168 h corrosion sample, which is shown in Figure 8c for the protruding dark-carbide phase at the coating/molten Zn interface. Therefore, the fabricated bushing surface coating has a uniform corrosion mechanism and its corrosion rate is predictable for the sink-roller assembly in continuous galvanised Zn-coating production applications.

### 3.3. Corrosion of Sleeve-Pipe Surface Coatings

The HVOF sleeve-pipe surface coating after laser treatment has a thickness of 1 mm. Their thicknesses after corrosion are 0.8 mm for 72 h, 0.6 mm for 120 h and 0.3 mm for 168 h, which leads to a coating corrosion rate of 0.0042 mm/h. This can be divided into a corrosion rate of 0.0033 mm/h for the first 120 h and 0.0063 mm/h for the last 48 h from 120 h to 168 h in experiments. The corrosion rate of the HVOF sleeve-pipe surface coating without laser treatment is 0.104 mm/h for the first 72 h, with the failure of the coatings occurring after 120 h from the previous experiment.

The samples after corrosion at various times for the sleeve-pipe coatings at 72 h, 120 h and 168 h are shown in Figure 9a–c, respectively.

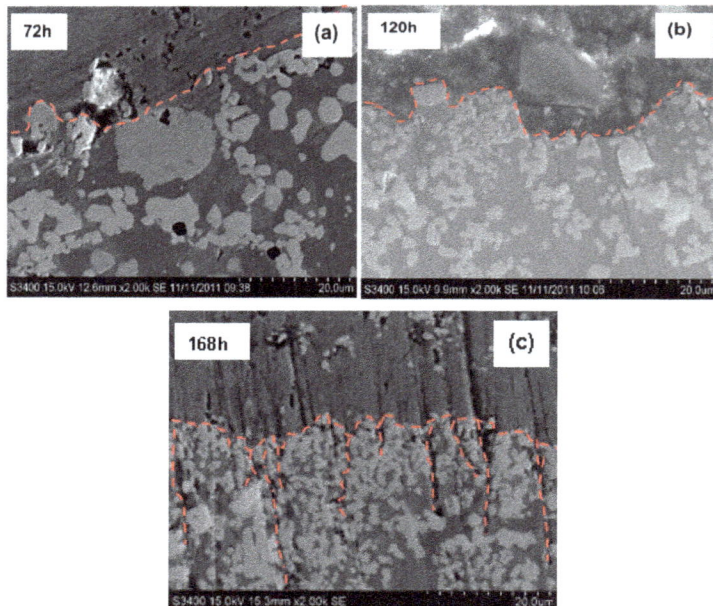

**Figure 9.** Microstructure of the sleeve coating after corrosion tests at: (**a**) 72 h; (**b**) 120 h; (**c**) 168 h.

Similar to the bushing coatings, the 72 h corrosion experiment (Figure 9a) also shows two regions: (a) the top left region with the presence of molten Zn on the coating surface and (b) the right bottom region with the coating. At this stage of the experiment, the coating morphology does not change, no new phase is formed and no corrosion channels or cracks are formed in the coating. It appears that the large carbide particles may be more corrosive-resistant to the molten Zn due to the particulates protruding more at the coating/molten Zn interface. However, no obvious corrosion rate difference exists between the matrix alloy phase (slightly dark-coloured) and the white-coloured particulate phases. That can still classify it as a selective phase corrosion mechanism at this stage of corrosion for the HVOF-coated sleeve surface.

Figure 9b shows the SEM image of the sample after 120 h of corrosion. There is clear evidence that the coating exhibited severe molten Zn corrosion at this stage of the experiment. Both the top matrix phase and large carbide particulates were corroded away from the coating into the molten Zn. This resulted in smaller carbide particle particulates, which were not subjected to laser treatment, and the matrix coatings further towards the substrate being left behind. In addition, a noticeable higher corrosion rate of the matrix alloy phase was found when compared to the white carbide particulate phase. However, the preferred corrosion resistance of the carbide particles did not result in dramatic changes in the coating corrosion rate or the formation of cracks or corrosion channels in the coating. This preferred corrosion difference is obvious in the 168 h corrosion sample. This is shown in Figure 9c, where white carbide particulates protrude with severe cracks and channels having opened in the coating to expose the severe corrosion of the matrix phase. Following this, the corrosion rates are obviously higher for the matrix phase when compared to the white carbide particulates. In Co-Cr-Ni alloy coatings, the corrosion begins from the anodic microstructural phase. If the form of the anodic phase is continuous, such as a grain-boundary coating structure, a rapid penetration of the component could occur.

It should be emphasised here that the current report only involved the report of bushing and sleeve coating performance in molten Zn without subjecting an applied force. During the actual sink-roller operation in a production environment, conditions are much more aggressive and complicated, as the bushing and sleeve pipe will be subjected to molten Zn corrosion and surface wear. The coating wear mechanisms are quite complex. At the initial stage of coating wear, there will be two sliding surfaces with the liquid Zn acting as a lubricant and corrosive agent. As time goes on, corrosion, such as Co-Zn, Cr-Zn and surface wear debris (such as metals or alloys of Cr, Co and hard-phase carbides, including WC, Cr-C and $\eta$-Co-W-C), will be added to the surface, which further complicates the wear-and-tear process of the surface coating. We are now trying to improve the coating qualities and to plan the wear performance of the coating in both simulated and actual production environments. The results will be subsequently published.

## 4. Conclusions

- The HVOF-coated sleeve-pipe sample exhibits a higher corrosive resistance of 0.0042 mm/h, when compared to the corrosive resistance (0.008 mm/h) of the welding-overlaid coated bushing sample. However, at the later stage of the corrosion, a preferred corrosion rate resulted, which caused cracks and corrosion channels in the coating. These are unfavourable for predicting the corrosion rate of components in applications.
- The slightly dark-coloured matrix material consisted of metal-based elements, such as Co and Cr. These elements are reactive to molten Zn and form corrosive products, such as Co-Zn or Cr-Zn. However, alloyed metals or carbide particles are usually inert to liquid Zn.
- The laser-treated top layer surface has more corrosive resistance than the untreated bottom layer of the sleeve surface coatings. This reveals that the laser treatment not only caused carbide particle growth, but also resulted in alloying of the matrix elements. This is the probable reason for its ability to also detect W and C in the dark-matrix phase in the EDS analysis. This dark-matrix phase is probably the dissolution of the WC phase into the matrix phase, which forms $\eta$-Co-W-C, a well-known Zn-corrosion resistance phase.

**Acknowledgments:** Thanks Xiao Dongshan from IFMTECH Co., Ltd. (Chongqing, China) for giving guidance and support at the experimental stage, providing so much valuable advice for the experiment. This work was funded by the National Natural Science Foundation of China (No. 51301112), Natural Science Foundation of Liaoning Province of China (No. 201602553).

**Author Contributions:** For this research article, G.Z. performed the experiments and wrote the manuscript; D.L. designed the experimental process; N.Z. analyzed the data; N.N.Z. assisted in paper review; S.D. reviewed the data.

**Conflicts of Interest:** The authors declare no conflict of interest.

## References

1.   Yue, Q.; Zhang, C.; Xu, Y.; Zhou, L.; Kong, H.; Wang, J. Performance of flow and heat transfer in a hot-dip round coreless galvanizing bath. *Metall. Mater. Trans. B* **2007**, *48*, 1188–1199. [CrossRef]

2.   Wang, H.; Zhang, G.; Meng, X.; Duan, S.; Xu, C.; Guan, S. Zinc corrosion resistance and inactivation process of sink roll in hot galvanization. *Chin. Surf. Eng.* **2012**, *25*, 42–46.

3.   Abro, M.A.; Lee, D.B. Microstructural Changes of Al Hot-Dipped P91 Steel during High-Temperature Oxidation. *Coatings* **2017**, *7*, 31. [CrossRef]

4.   Matthews, S.; James, B. Review of thermal spray coating applications in the steel industry: Part 2–Zinc pot hardware in the continuous galvanizing line. *Therm. Spray Technol.* **2010**, *19*, 1277–1286. [CrossRef]

5.   Reichinger, M.; Bremser, W.; Dornbusch, M. Interface and volume transport on technical cataphoretic painting: A comparison of steel, hot-dip galvanised steel and aluminium alloy. *Electrochim. Acta* **2017**, *231*, 135–152. [CrossRef]

6.   Zhang, K.; Battiston, L. Friction and wear characterization of some cobalt- and iron-based superalloys in zinc alloy baths. *Wear* **2002**, *252*, 332–344. [CrossRef]

7.   Zhang, L. Thermal Spray Fe-Al Coatings Using for Continuous Hot Galvanizing Sink Rolls. Master's Thesis, Dalian Maritime University, Dalian, China, 5 March 2007.

8.   Wang, L.; Zhou, Y.; Chen, G.; Rao, S. Analysis on dominant influencing factors of on-line life cycle in sink roll system and effective improved methods. *Eng. Fail. Anal.* **2015**, *58*, 8–18. [CrossRef]

9.   Guelton, N.; Lopès, C.; Sordini, H. Cross Coating Weight Control by Electromagnetic Strip Stabilization at the Continuous Galvanizing Line of ArcelorMittal Florange. *Metall. Mater. Trans. B* **2016**, *47*, 2666–2680. [CrossRef]

10.  Bae, K.T.; La, J.H.; Lee, I.G.; Lee, S.Y. Effects of annealing heat treatment on the corrosion resistance of Zn/Mg/Zn multilayer coatings. *Met. Mater. Int.* **2017**, *23*, 481–487. [CrossRef]

11.  Zhang, J.; Deng, C.; Song, J.; Zhou, K. MoB-CoCr as alternatives to WC-12Co for stainless steel protective coating and its corrosion behavior in molten zinc. *Surf. Coat. Technol.* **2013**, *235*, 811–818. [CrossRef]

12.  Tsipas, D.N.; Triantafyllidis, G.K.; Kiplagat, J.K.; Psillakia, P. Degradation behavior of boronized carbon and high alloy steels in molten aluminium and zinc. *Mater. Lett.* **1998**, *37*, 128–131. [CrossRef]

13.  Bahadormanesh, B.; Ghorbani, M.; Kordkolaei, N.L. Electrodeposition of nanocrystalline Zn/Ni multilayer coatings from single bath: Influences of deposition current densities and number of layers on characteristics of deposits. *Appl. Surf. Sci.* **2017**, *404*, 101–109. [CrossRef]

14.  Guo, J.; Zhang, C.; Ji, X.; Wan, W. Properties of coating by nano WC-Co used on sink roll. *Therm. Spray Technol.* **2016**, *8*, 25–29.

*coatings*

MDPI

*Article*

# The Use of Triboemission Imaging and Charge Measurements to Study DLC Coating Failure

**Alessandra Ciniero [1,\*], Julian Le Rouzic [2] and Tom Reddyhoff [1,\*]**

[1]  Tribology Group, Department of Mechanical Engineering, Imperial College London, London SW7 2AZ, UK
[2]  Institut P', CNRS, Université de Poitiers, ISAE-ENSMA, F-86962 Futuroscope Chasseneuil, France;
      julian.le.rouzic@univ-poitiers.fr
\*  Correspondence: alessandra.ciniero11@imperial.ac.uk (A.C.); t.reddyhoff@imperial.ac.uk (T.R.);
      Tel.: +44-207-5943840 (A.C. & T.R.)

Received: 26 July 2017; Accepted: 17 August 2017; Published: 20 August 2017

**Abstract:** We present a study on the simultaneous evolution of the electron emission and surface charge accumulation that occurs during scratching tests in order to monitor coating failure. Steel discs coated with a diamond-like-carbon (DLC) film were scratched in both vacuum (~$10^{-5}$ Torr) and atmospheric conditions, with electron emission and surface charge being measured by a system of microchannel plates and an electrometer, respectively. The results highlight a positive correlation between emission intensity values, surface charge measurements and surface damage topography, suggesting the effective use of these techniques to monitor coating wear in real time.

**Keywords:** triboemission; tribocharging; coating-failure; wear; diamond-like-carbon

## 1. Introduction

Coatings are used extensively in many industrial and commercial applications in order to protect components that are subjected to sliding and rolling contact, both with and without liquid lubricants [1]. It has been shown that hard coatings and other surface modification methods are able to improve the resistance of rolling elements (i.e., bearings) to friction, wear and corrosion [2,3]. Thin diamond-like-carbon (DLC) coatings, for instance, provide protective, low friction, wear resistant surfaces for numerous industrial applications such as invasive and implantable medical devices, razor blades, magnetic hard discs and microelectromechanical systems [4–10]. However, because of mismatches in mechanical and electrical properties between the coating and substrate, residual stresses and chemical reactions can arise at the interface of the materials leading to the structural degradation of the coating. The evaluation of damage and the electrical state of the film/coating and substrate, along with the option to monitor the coatings failure are becoming important tools in the endeavour to extend the lifetime of widely used devices.

Recently, several techniques have been used to investigate coating failure. These include atomic force microscopy to map the evolution of cracks in nickel films on a polyimide substrate [11]; ultrasonic force microscopy to measure the debonding of glass films on polyethylene terephthalate substrate [12]; and scanning electron microscopes and thermographs to detect damage evolution [13,14]. In addition to these qualitative techniques, real-time in situ methods based on acoustic emission have been developed to obtain quantitative stress/strain information to study interfacial properties of coating/film systems [15,16].

In this paper, we propose a new in situ and real-time technique to monitor the failure of diamond-like-carbon coating on a steel substrate during sliding tests under vacuum and atmospheric conditions. The method combines triboemission imaging and tribocharging measurements. Triboemission refers to the emission of charged particles such as electrons, protons, positive ions and negative ions that occur during surface damage (i.e., cracking formation, wear) [17–20]—see the supplementary example video showing imaging of continuous electron emission arising from a moving alumina specimen scratched by a stationary diamond tip. These emissions correlate positively with the electrical resistivity of the rubbed material, which decreases from insulators to conductors (in the order: insulating > semi-conductive > conductive) [17,21]. Furthermore, our recent studies on the spatial characteristics of triboemission bursts have shown that their direction, shape, size and intensity depend on the failure mode of the materials (such as cracking and grain pull-out) [17,22]. Tribocharging measurements, on the other hand, are used to monitor the variation of the charge on the surface, which may be related to tribochemical reactions occurring at the surface contact area [23,24].

This work focuses on correlating the evolution of surface topography with the variation in emission intensity and the measured charge to demonstrate the applicability of electron emission and surface charge measurements as a means of studying coating failure. Finally, a comparison is made between charge measurements, obtained in vacuum and atmosphere, in order to show that such measurements are an effective means of detecting coating failure in practice.

## 2. Materials and Methods

Discs, made from 52100 steel, coated with a 1 μm layer of DLC (a-C: H $sp^3$~50%, H~40%–characterisation provided by PCS Instruments, London, UK) were used as test specimens. These discs had a diameter of 46 mm and a thickness of 6 mm and were cleaned with toluene followed by isopropanol in an ultrasonic bath (15 min for each chemical), prior to each test.

The tribometer used to conduct the study is represented schematically in Figure 1 [17,22]. It consists of a system of microchannel plates (MCPs, i.e., arrays of electrons multipliers), coupled with a phosphor screen (Photonis Inc., Sturbridge, MA, USA). The two circular MCPs, in a chevron arrangement with an active diameter of 75 mm, are located 10 mm above and parallel to the disc specimen (note: the centre of the circular MCPs are located directly above the centre of rotation of the disc). This setup allows us to obtain spatially resolved images of the triboemission, with 1:30 magnification due by the divergence of emitted electrons. The emissions detected and visualised through this system were recorded by a high speed camera (Phantom Miro eX, Vision Research Ltd., Bedford, UK) with a Fujian 35 mm f1.7 lens located above the experimental setup. The sliding contact was produced by loading a diamond tip of radius 100 μm (Synton-MDP Ltd., Port, Switzerland) against the rotating disc specimen. The rotation of the disc was recorded by the supplied PCS Instruments encoder device. In addition, an electrometer (Model: 6517b, Keithley Instruments Ltd., Bracknell, UK) coupled with a 10 mm × 5 mm metal sheet electrode attached underneath the specimen was used to inductively measure the charge of the surface, simultaneously with the emission detection. The tests were conducted in vacuum conditions at a pressure of ~$10^{-5}$ Torr.

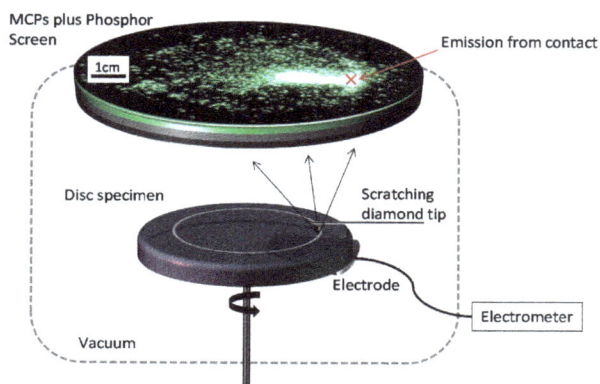

**Figure 1.** Schematic of the tribometer.

The triboemission measurements were conducted in negative particle detection mode, i.e., 10%–85% of 0.01–50 keV electrons were detected [25], with the voltage applied to the input MCP, output MCP and the phosphor screen being ground, applying 1.5 kV, and 5 kV, respectively. The speed of the rotation of the disc was 4 Hz giving a sliding velocity of 50 mm/s. The deadweight load applied to the scratching tip on to the disc was 0.5 N. The frame rate of the high-speed camera (exposure time 8 ms), the encoder and the electrometer acquisition were synchronised at 125 Hz. At the end of the test, the cleaning procedure previously described was repeated and the topography of the wear track was recorded using the Veeco Wyko NT9100 optical profiler (Veeco Instruments Inc., Plainview, NY, USA).

The failure of the coating in atmospheric conditions was assessed by focusing the high speed camera directly on the outlet of the contact. In this case, the MCPs and the phosphor screen system were replaced by the 5× magnification lenses.

## 3. Results and Discussion

### 3.1. Triboemission Measurements

The test apparatus used here differs significantly from those used in previous studies that measured the triboemission from hydrogenated carbon films in the vicinity of a sliding contact [26–30]. The spatial resolution achieved with this technique allows the emission to be visualised, defining its shape and size. This provides more detailed information than was previously obtained from single point measurements.

Figure 2a displays the average (spatial) intensity of each phosphor screen image plotted as a function of time, throughout the entire sliding test. In addition, Figure 2b shows an example of an emission event as viewed on the phosphor screen. The emission event is localised at the tip location suggesting that it is due to wear. In Figure 2a, a decrease in emission intensity is also evident during the second half of the test.

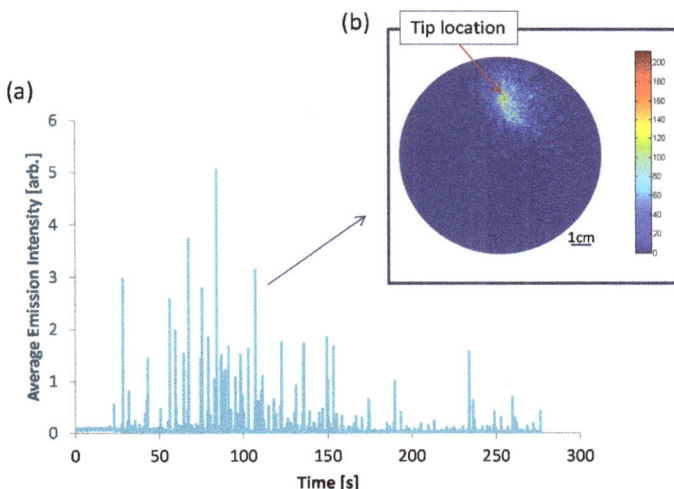

**Figure 2.** (a) Average Emission Intensity (spatial average of each phosphor screen image) vs. time, (complete test); (b) Phosphor screen intensity of a single emission event.

An alternative way to display the same phosphor screen intensity data is to plot the average emission intensity as a function of the cycles and the angular disc position of the stylus relative to the disc, as shown in Figure 3 (with each coloured square representing the average phosphor screen intensity). Here, the spatial evolution of the emissions can be visualised and divided into three regions.

Initially, there is a moderate level of emission indicated by the pale blue rectangular section at the start of the test. Towards the end of this period, the first and highest intensity peak in emission occurs, and the low-level baseline emission transitions into the second stage of the test (shown by the background colour becoming dark blue). From this point onwards, sporadic high intensity peaks occur, although their intensity is lower than that of the first event. Eventually, the frequency and magnitude of these peaks reduce, as shown by the mostly uninterrupted dark blue region.

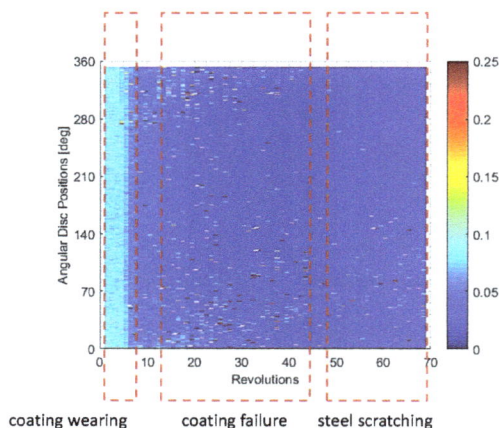

**Figure 3.** Average phosphor screen intensity as a function of time and disc position.

It is hypothesised that this observed behaviour is attributed to the following mechanism. The initial area characterised by a constant, low level intensity (the light blue colour) defines the emission generated by the wearing of the non-conductive coating layer as shown by the optical microscopy scan in Figure 4a. This is in accordance with previous studies on non-conductive materials [17]. The following high intensity emission event indicates the moment at which the coating fails resulting in a wear trace characterised by both surface cracks and partial delamination (Figure 4b). This is in accordance with previous studies that suggest that emission is due to the high energy generated during the damage of the surface, in particular crack formation [17,19,31,32]. The third region shows relatively few high emission events, occurring due to wearing of the remaining coating. The baseline emission (shown by the dark blue colour) is at a low level due to the scratching of the metal substrate, which can conduct away charge and hence prevents emission. This is in agreement with the fact that the triboemission intensity is greater for materials with high electrical resistivity [33].

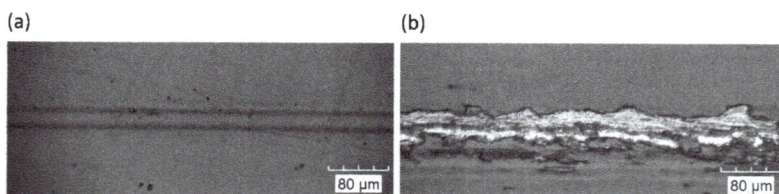

**Figure 4.** Optical Microscopy scans of the wear trace during (**a**) coating wearing phase; (**b**) coating failure phase.

Since the angular position of the disc is recorded and synchronised with the camera acquisition, the average emission intensity can be plotted as a function of the angular position of the disc as shown in Figure 5. Here, it is evident that the highest intensity emissions are localised around 90° and the lowest intensity emission are around 270° of disc rotation. Visual inspection of the disc revealed that the depth of wear around the track was not uniform. It was therefore hypothesized that the area of the disc characterised by high intensity emission events corresponded to the diamond tip contacting the non-conductive coating. Conversely, the area characterised by low emission events corresponded with the tip scratching the conductive steel.

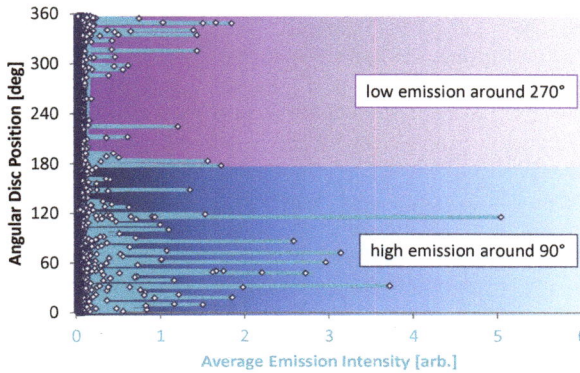

**Figure 5.** Average of phosphor screen intensity as a function of emission intensity and disc position.

After the test, a Veeco optical profilometer analysis of the surface was conducted to evaluate the depth of the wear track in the two areas under consideration. Results are shown in Figure 6. It can be seen that the wear track at the angular position of 270° is 1.69 μm, which is deeper than the thickness of the coating layer (Figure 6b). In the region of the disc surface around the 90° location, the depth is only 0.1042 μm, which is less than the thickness of the coating layer, as shown in Figure 6d. This proves that, around the 90° location, the coating was not removed so that the tip scratched the coating generating an emission in accordance with triboemission measurements for non-conductive materials. However, around 270°, the tip scratched the metal substrate and the intensity of the emission was reduced in accordance with previous emission measurements for conductive materials.

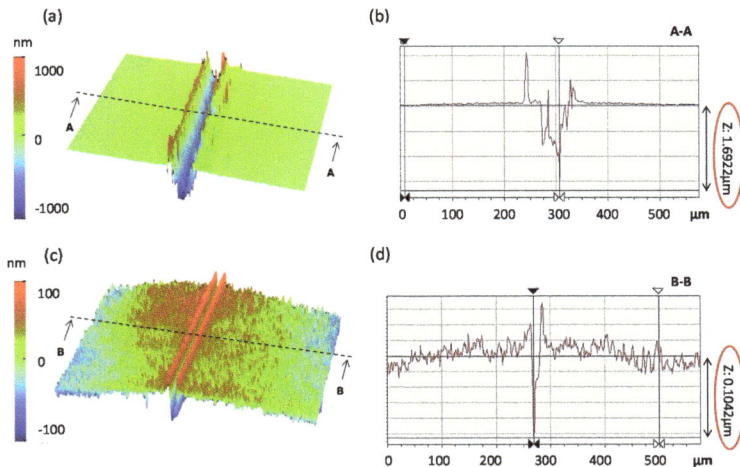

**Figure 6.** Veeco analysis of the surface at disc regions of low and high emission, (**a**) 3D profile at 270°; (**b**) section depth at 270°; (**c**) 3D profile at 90°; (**d**) section depth at 90°.

These results clearly show the possibility of using this technique to monitor the coating failure in real time by evaluating the evolution of the emission events and by identifying the area of the failure. There is an issue however, which is that the microchannel plate measurements must be carried out within a vacuum. For this reason, the following charge measurements were performed.

## 3.2. Charge Measurements

The release of the emission was then compared with the evolution of the charge on the disc surface. To do this, instantaneous measurements of charge on the surface and the average of the emission intensity are plotted as a function of time in Figure 7. There is a clear correspondence between the emission intensity and the charge accumulation for the DLC surface (a local regression smoothing method was applied to the charge measurement data for a better comparison). In the first part of the test, there is a clear correspondence between the negative emission peaks and the positive charge peaks. This suggests that the surface charges positively, at least partly due to electrons leaving the surface. At around 100 s, the reduction of the emission intensity and the charge can be attributed to the failure of the coating. It is suggested that this reduction is caused by: (i) the scratching of a reduced area of coating (part of the coating is removed during the first 100 s of the test); (ii) the exposure of the metal substrate which switches the contact from being diamond/DLC to being diamond/steel, preventing the accumulation of opposite charge on each of the two counter bodies which would otherwise lead to emission; and (iii) the exposure of the steel substrate which may absorb charge, preventing the generation of emission.

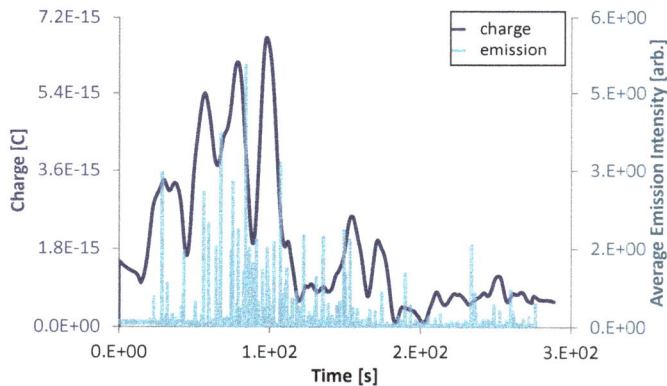

**Figure 7.** Instantaneous values of average phosphor screen intensity and surface charge vs. time.

Additional measurements were assessed to compare the differences in charging behaviour between tests in which coating failure either did or did not occur (Figure 8). The curves of instantaneous charge values show a fluctuation for both no failure and failure tests—see Figure 8a,c, respectively. Figure 8a shows that, in the case of no coating failure, the charge fluctuates around 10 pC for the entire test. However, the cumulative values in Figure 8b show positive charges accumulating on the surface, as also reported by previous studies [34], until a saturation value is reached, from which point onwards the charging remains stable. In the case of coating failure, the instantaneous charge curve fluctuates around 10 pC until the point failure, after which it increases, as shown in Figure 8c. The cumulative curve in Figure 8d shows that, after failure, the surface started to accumulate negative charge. This is suggested to be caused by the reasons outlined above.

This comparison clearly shows the differing charging behaviours in the case of failure and no failure of the coating. The accumulation on the surface of positive particles due to the emission of negative particles is affected in the case of failure.

(a)

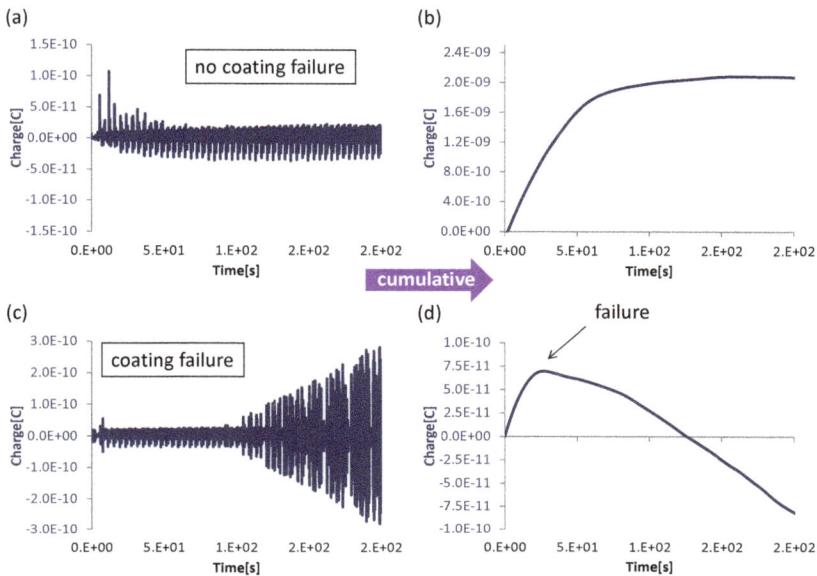

(b)

(c)

(d)

**Figure 8.** Surface charge vs. time: (**a**) no coating failure—instantaneous; (**b**) no coating failure—cumulative; (**c**) coating failure—instantaneous; (**d**) coating failure—cumulative.

### 3.3. Measurements under Atmosphere

The final investigation was aimed at comparing the charging of the surface in both vacuum and atmospheric conditions. A comparison for the charge measurements when failure did and did not occur is reported in Figure 9. When the coating remains intact, the surface charges positively in both vacuum and atmospheric conditions, as shown in Figure 9a. In the case of failure, the surface charges negatively as soon as the metal substrate is exposed, as reported in Figure 9b. In each case, the value of the charge in the atmosphere is lower compared to that under vacuum conditions. This may be due to increased electrical breakdown due to the presence of air. Specifically, when the electric field due to the accumulation of charge on components exceeds a threshold determined by the dielectric field strength [35] of the surrounding gas (30 kV/cm for air [36]), breakdown can occur through a Townsend discharge process. Furthermore, some evidence suggests that corona discharge limits the formation of charge during contact electrification [37,38].

(a)

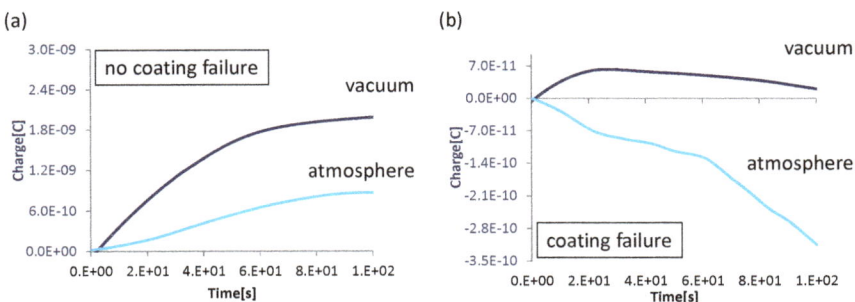

(b)

**Figure 9.** Surface charge vs time in vacuum and atmosphere: (**a**) no coating failure—cumulative; (**b**) coating failure—cumulative.

The charge measurements in atmosphere were accompanied by standard optical microscope video recordings of the wear track. This gave an in situ indication of the instantaneous coating integrity on the specimen surface during sliding, by acquiring and averaging the colour intensity of pixels recorded at the exit of the contact. The increase in the intensity indicates the exposure of the metal substrate, since steel has a higher reflectivity compared to DLC coating. The average of the intensity at the exit of the contact and the measured surface charge are plotted against time in Figure 10a. This plot shows that, when the tip wears the coating without failure, the surface charges positively. However, as soon as the coating fails (the moment of the failure is indicated by the dashed line) the metal substrate is exposed, as show in Figure 10b, and the surface charge begins to decrease towards negative values in accordance with previous results.

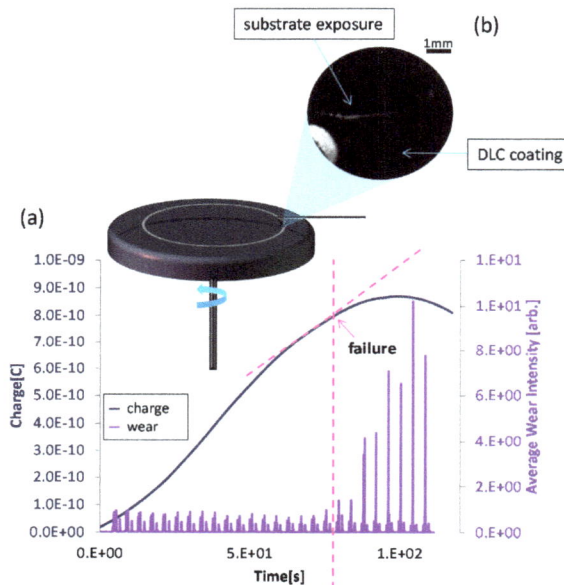

**Figure 10.** (a) Average wear track intensity and surface charge vs. time; (b) example of 5× magnification wear track tip outlet.

## 4. Conclusions

The present paper reports the potential of measuring charged particle emission and/or tribocharging for studying or monitoring the failure of coatings. The spatial information of triboemission allows us to identify the exact time and location of the failure. The evolution of the failure during the sliding and the change in the characteristic of the contact can also be monitored. The phosphor screen images show that the onset of coating failure is accompanied by high intensity emission events, while the exposure of the metal substrate is shown to cause a decrease in background emission intensity. This is attributed to the conductive substrate preventing the build-up of charge. The comparison between the emission and the charging of the surface again shows that triboemission is linked to accumulation of the charge on the surface. Specifically, the peaks of emission and peaks of positive charge coincide, showing that the surface charge is influenced by the negative particle emission leaving the contact.

In atmosphere, the trends in the charge measurements are comparable with those carried out in vacuum, but their magnitudes are consistently lower. The surface charges positively during sliding contact until failure occurs, after which, it charges negatively. It is suggested that the values measured in atmosphere are lower than in vacuum either due to the oxidation reactions which passivate the active fresh material or because the presence of air leads to dielectric breakdown. The charge measurements are supported by qualitative measurements of surface reflectivity. Here, the appearance of the conductive metal substrate is followed by a change from positive to negative surface charge. Overall, these results suggest that charge and emission measurements may be effective in monitoring the failure of coatings in real time and their ability to provide more detailed information will be assessed in future work.

**Supplementary Materials:** The supplementary video, Video S1: Electron emission from alumina specimen, is available online at http://www.mdpi.com/2079-6412/7/8/129/s1.

**Acknowledgments:** This research was supported by the UK Engineering and Physical Research Sciences Research Council, with equipment funding provided by the Taiho Kogyo Tribology Research Foundation.

**Author Contributions:** A.C. and T.R. conceived and designed the experiments; A.C. performed the experiments and analyzed the data; J.L.R. contributed for the design of the test rig; A.C. wrote the paper.

**Conflicts of Interest:** The authors declare no conflict of interest.

## References

1. Holmberg, K.; Mathews, A. Coatings tribology: a concept, critical aspects and future directions. *Thin Solid Films* **1994**, *253*, 173–178. [CrossRef]
2. Stewart, S.; Ahmed, R. Rolling contact fatigue of surface coatings—A review. *Wear* **2002**, *253*, 1132–1144. [CrossRef]
3. Maurer, R. Friction, wear, and corrosion control in rolling bearings through coatings and surface modification: A review. *J. Vac. Sci. Technol. A* **1986**, *4*, 3002–3006. [CrossRef]
4. Raveh, A.; Martinu, L.; Gujrathi, S.C.; Klemberg-Sapieha, J.E.; Wertheimer, M.R. Structure-property relationships in dual-frequency plasma deposited hard aC: H films. *Surf. Coat. Technol.* **1992**, *53*, 275–282. [CrossRef]
5. Raveh, A.; Martinu, L.; Hawthorne, H.M.; Wertheimer, M.R. Mechanical and tribological properties of dual-frequency plasma-deposited diamond-like carbon. *Surf. Coat. Technol.* **1993**, *58*, 45–55. [CrossRef]
6. Martinu, L. Hard carbon films deposited under high ion flux. *Thin Solid Films* **1992**, *208*, 42–47. [CrossRef]
7. Butter, R.; Allen, M.; Chandra, L.; Lettington, A.H.; Rushton, N. In vitro studies of DLC coatings with silicon intermediate layer. *Diam. Relat. Mater.* **1995**, *4*, 857–861. [CrossRef]
8. Snyders, R.; Bousser, E.; Amireault, P.; Klemberg-Sapieha, J.E.; Park, E.; Taylor, K.; Casey, K.; Martinu, L. Tribo-Mechanical Properties of DLC Coatings Deposited on Nitrided Biomedical Stainless Steel. *Plasma Process. Polym.* **2007**, *4*, S640–S646. [CrossRef]
9. Grill, A. Tribology of diamondlike carbon and related materials: An updated review. *Surf. Coat. Technol.* **1997**, *94*, 507–513. [CrossRef]
10. Grill, A. Diamond-like carbon: state of the art. *Diam. Relat. Mater.* **1999**, *8*, 428–434. [CrossRef]
11. George, M.; Coupeau, C.; Colin, J.; Grilhé, J. Atomic force microscopy observations of successive damaging mechanisms of thin films on substrates under tensile stress. *Thin Solid Films* **2003**, *429*, 267–272. [CrossRef]
12. McGuigan, A.P.; Huey, B.D.; Briggs, G.A.D.; Kolosov, O.V.; Tsukahara, Y.; Yanaka, M. Measurement of debonding in cracked nanocomposite films by ultrasonic force microscopy. *Appl. Phys. Lett.* **2002**, *80*, 1180–1182. [CrossRef]
13. Qian, L.; Zhu, S.; Kagawa, Y.; Kubo, T. Tensile damage evolution behavior in plasma-sprayed thermal barrier coating system. *Surf. Coat. Technol.* **2003**, *173*, 178–184. [CrossRef]
14. Busso, E.P.; Wright, L.; Evans, H.E.; McCartney, L.N.; Saunders, S.R.J.; Osgerby, S.; Nunn, J. A physics-based life prediction methodology for thermal barrier coating systems. *Acta Mater.* **2007**, *55*, 1491–1503. [CrossRef]
15. Lu, P.; Chou, Y.K.; Thompson, R.G. Short-time Fourier Transform Method in AE Signal Analysis for Diamond Coating Failure Monitoring in Machining Applications. In Proceedings of the ASME 2010 International Manufacturing Science and Engineering Conference, Erie, PA, USA, 12–15 October 2010; Volume 1.

16. Mao, W.G.; Wu, D.J.; Yao, W.B.; Zhou, M.; Lu, C. Multiscale monitoring of interface failure of brittle coating/ductile substrate systems: A non-destructive evaluation method combined digital image correlation with acoustic emission. *J. Appl. Phys.* **2011**, *110*, 084903. [CrossRef]
17. Ciniero, A.; Le Rouzic, J.; Baikie, I.; Reddyhoff, T. The Origins of Triboemission—Correlating Electron Emission with Surface Damage. *Wear* **2017**, *374–375*, 113–119. [CrossRef]
18. Dickinson, J.; Donaldson, E.; Park, M. The emission of electrons and positive ions from fracture of materials. *J. Mater. Sci.* **1981**, *16*, 2897–2908. [CrossRef]
19. Nakayama, K.; Suzuki, N.; Hashimoto, H. Triboemission of charged particles and photons from solid surfaces during frictional damage. *J. Phys. D Appl. Phys.* **1992**, *25*, 303–308. [CrossRef]
20. Molina, G.J.; Furey, M.J.; Ritter, A.L.; Kajdas, C. Triboemission from alumina, single crystal sapphire, and aluminum. *Wear* **2001**, *249*, 214–219. [CrossRef]
21. Nakayama, K.; Hashimoto, H. Triboemission from various materials in atmosphere. *Wear* **1991**, *147*, 335–343. [CrossRef]
22. Le Rouzic, J.; Reddyhoff, T. Spatially Resolved Triboemission Measurements. *Tribol. Lett.* **2014**, *55*, 245–252. [CrossRef]
23. Nakayama, K.; Nevshupa, R.A. Plasma generation in a gap around a sliding contact. *J. Phys. D Appl. Phys.* **2002**, *35*, L53–L56. [CrossRef]
24. Nakayama, K.; Nevshupa, R.A. Effect of dry air pressure on characteristics and patterns of tribomicroplasma. *Vacuum* **2004**, *74*, 11–17. [CrossRef]
25. Wiza, J.L. Microchannel plate detectors. *Nucl. Instrum. Methods* **1979**, *162*, 587–601. [CrossRef]
26. Matta, C.; Eryilmaz, O.L.; De Barros Bouchet, M.I.; Erdemir, A.; Martin, J.M.; Nakayama, K. On the possible role of triboplasma in friction and wear of diamond-like carbon films in hydrogen-containing environments. *J. Phys. D Appl. Phys.* **2009**, *42*, 075307. [CrossRef]
27. Nakayama, K.; Yamanaka, K.; Ikeda, H.; Sato, T. Friction, wear, and triboelectron emission of hydrogenated amorphous carbon films. *Tribol. Trans.* **1997**, *40*, 507–513. [CrossRef]
28. Nakayama, K.; Ikeda, H. Triboemission characteristics of electrons during wear of amorphous carbon and hydrogenated amorphous carbon films in a dry air atmosphere. *Wear* **1996**, *198*, 71–76. [CrossRef]
29. Nakayama, K. Triboemission of electrons, ions, and photons from diamondlike carbon films and generation of tribomicroplasma. *Surf. Coat. Technol.* **2004**, *188–189*, 599–604. [CrossRef]
30. Nakayama, K.; Bou-Said, B.; Ikeda, H. Tribo-Electromagnetic Phenomena of Hydrogenated Carbon Films—Tribo-Electrons, -Ions, -Photons, and -Charging. *J. Tribol.* **1997**, *119*, 764–768. [CrossRef]
31. Dickinson, J.T. Fracto-emission: The role of charge separation. *J. Vac. Sci. Technol. A* **1984**, *2*, 1112–1116. [CrossRef]
32. Walton, A.J. Triboluminescence. *Adv. Phys.* **2006**, *26*, 887–948. [CrossRef]
33. Nakayama, K. Tribocharging and friction in insulators in ambient air. *Wear* **1996**, *194*, 185–189. [CrossRef]
34. Kornfeld, M. Frictional electrification. *J. Phys. D Appl. Phys.* **1976**, *9*, 1183–1192. [CrossRef]
35. Vella, S.J.; Chen, X.; Thomas, S.W., III; Zhao, X.; Suo, Z.; Whitesides, G.M. The determination of the location of contact Electrification-induced discharge Events. *J. Phys. Chem. C* **2010**, *114*, 20885–20895. [CrossRef]
36. Lide, D.R.; Haynes, W.M.; Bruno, T.J. *CRC Handbook of Chemistry and Physics*, 9th ed.; CRC Press: Boca Raton, FL, USA, 2015.
37. Fabian, A.; Krauss, C.; Sickafoose, A.; Horanyi, M.; Robertson, S. Measurements of electrical discharges in Martian regolith simulant. *IEEE Trans. Plasma Sci.* **2001**, *29*, 288–291. [CrossRef]
38. Thomas, S.W.; Vella, S.J.; Kaufman, G.K.; Whitesides, G.M. Patterns of electrostatic charge and discharge in contact electrification. *Angew. Chem.* **2008**, *120*, 6756–6758. [CrossRef]

*coatings*

MDPI

*Article*

# The Phase Evolution and Property of FeCoCrNiAlTi$_x$ High-Entropy Alloying Coatings on Q253 via Laser Cladding

**Bin He [1,2], Nannan Zhang [1,*], Danyang Lin [1], Yue Zhang [1], Fuyu Dong [1] and Deyuan Li [1]**

[1]    Department of Material Science and Engineering, Shenyang University of Technology, Shenyang 110870, China; hebin2015021@sut.edu.cn (B.H.); lindanyang2014136@sut.edu.cn (D.L.); yuezhang@sut.edu.cn (Y.Z.); dongfuyu@sut.edu.cn (F.D.); dmy1962@sut.edu.cn (D.L.)

[2]    Department of Pipeline, Shenyang Institute of Special Equipment Inspection & Research, Shenyang 110032, China

*    Correspondence: zhangnn@sut.edu.cn; Tel.: +86-24-2549-6812

Received: 18 August 2017; Accepted: 25 September 2017; Published: 28 September 2017

**Abstract:** High-entropy alloys (HEAs) are emerging as a hot research frontier in the metallic materials field. The study on the effect of alloying elements on the structure and properties of HEAs may contribute to the progress of the research and accelerate the application in actual production. FeCoCrNiAlTi$_x$ ($x = 0$, 0.25, 0.5, 0.75, and 1 in at.%, respectively) HEA coatings with different Ti concentrations were produced on Q235 steel via laser cladding. The constituent phases, microstructure, hardness, and wear resistance of the coatings were investigated by XRD, SEM, microhardness tester and friction-wear tester, respectively. The results show that the structure of the coating is a eutectic microstructure of FCC and BCC$_1$ at $x = 0$. The structure of coatings consists of both proeutectic FCC phase and the eutectic structure of BCC$_1$ and BCC$_2$. With the continuous addition of Ti, the amount of eutectic structure decreases. The average hardness of the FeCoCrNiAlTi$_x$ HEA coatings at $x = 0$, 0.25, 0.5, 0.75, and 1 are 432.73 HV, 548.81 HV, 651.03 HV, 769.20 HV, and 966.29 HV, respectively. The hardness of coatings increases with the addition of Ti, where the maximum hardness is achieved for the HEA at $x = 1$. The wear resistance of the HEA coatings is enhanced with the addition of Ti, and the main worn mechanism is abrasive wear.

**Keywords:** laser cladding; high-entropy coatings; tribological property; phase evolution; FeCoCrNiAlTi$_x$

## 1. Introduction

The concept of high-entropy alloys (HEAs) is a new alloy design philosophy that was proposed recently, breaking the bottleneck stage of the conventional alloy design concept. The HEA philosophy considers that multiple principle elements in an alloy system will not produce brittle phases such as intermetallic compounds (IMCs) or other complex phases, leading to brittleness and difficulties in processing and application. On the contrary, the high entropy effect can maintain the simple solid solutions in alloys [1–3]. The main reason people think multiple principle elements will decrease the property of the alloy is the Gibbs phase rule. According to the Gibbs phase rule, $f = n - p + 1$ ($f$, freedom degree; $n$, component number; $p$, phase number), the equilibrium solidification phase number for $n$ kinds of elements is given by $p = n + 1$. Since phase formation processes are always non-equilibrium solidification, the number of phases tends to be given by $p > n + 1$. More precisely, as more types of principle elements are added to the alloy system, more complicated phases appear in it, which is harmful [4]. However, a new path of alloy design was carried out for the first time by Yeh et al. [5,6] in 2004. They found that there were no IMCs if more than five types of principle elements were added into the alloy system and simple solid solutions could be gained. More and more scholars have now joined in the study of HEAs. HEAs are better than traditional alloys in many aspects because of their high entropy effect especially in the preparation of wear and corrosion-resistant coatings, which are reported to be good candidates as structural and functional materials. Much research has been done on the properties of different HEAs. Duan et al. [7] found that the hardness of AlCoCrFeNiCu via arc melting was 475.3 HV, showing excellent tribological properties. Tian et al. [8] collected nearly 100 kinds of HEA experimental results and found that hardness steadily increased with different atomic sizes. Among the previous HEAs studied, HEAs are usually synthesized via arc melting technology or casting methods, which can result in high costs due to the amounts of expensive alloying elements that are added. As wear and corrosion resistance is a surface phenomenon and mainly determined by the surface properties of a material rather than by the bulk properties, laser cladding is widely adopted in the coating preparation since it has many advantages such as energy concentration, a low substrate dilution rate, and a fast cooling rate. Chen et al. [9] found that, with the increase in aluminum content, $Al_xCoFeNiCu_{1-x}$ HEAs displayed a greater hardness. An et al. [10] prepared $MoFeCrTiWAl_xSi_y$ coatings via laser cladding, which can obtain excellent quality coatings with the simultaneous addition of Si and Al.

Many HEA compositions are possible, and one well-known system is Al–Fe–Cr–Co–Ni. The evolution of the phase/microstructure was attributed to rapid quenching during laser processing. Alloy systems possess attractive properties such as high hardness, abrasiveness, and corrosion resistance. The addition of elements has a significant effect on the properties of HEA coatings [11,12]. It has been reported that Cu, Ni can promote the formation of a face-centered cubic (FCC) system, while Cr, Ti can lead to the formation of a body-centered cubic (BCC) system. The in-depth investigation on the effect of element addition on properties and structures may accelerate the application of HEA coatings. Thus, in this work, the effect of Ti addition on the properties and phase evolution of $FeCoCrNiAlTi_x$ coatings is investigated, and the results of the present study on the rules of phase evolution, microhardness, and wear resistance would provide theoretical guidance to further explore the effects on the HEAs of other element additions.

## 2. Experimental Procedure

In this experiment, Q235 steel (SD Steel, Jinan, China) was chosen as the substrate and the nominal chemical composition in wt % is 0.12%–0.2% C; 0.28% Cr; 0.2% Ni; 0.3% Si; 0.3%–0.67% Mn; 0.035% S; 0.04% P and Fe balance. The equiatomic ratio of the pure Co, Ni, Cr, and Al powder mixed with various concentrations of Ti powder (99.6%, 300 mesh, Tianjiu Ltd., Changsha, China) was chosen as the cladding powder. Fe powder was not included in the cladding powder since the Fe atoms can be introduced from the substrate due to the dilution, and the content can be regulated by controlling the power of laser. The mixed powder was ground via planetary ball mill for 10 h with a rotate speed of 200 r/min. The ground powder was preplaced on the surface of 100 mm × 100 mm × 10 mm cleaned Q235 substrates with a thickness of 1.5 mm. The laser generator (FL-Dlight-1500, Yuchen Ltd., Nanjing, China) was used to prepare the sample with a power of 800 W, a scanning speed of 3 mm/s, a spot size of 3 mm × 1 mm, an overlap rate of 30% with Ar protection (10 L/min), under which good quality $FeCoCrNiAlTi_x$ (x = 0, 0.25, 0.5, 0.75, and 1 in at.%, respectively) HEA coatings of moderate dilution could be achieved. The cladded samples were cut into small cross sections for metallography. The samples were ground using a grinding machine; grinding started at 100, and then proceeded at 200 and then 2000 abrasive papers, which was followed by polishing using a 1 m diamond slurry. They were then etched by a 5% nitric acid alcohol solution. The scanning electronic microscope (SEM, S3400, Hitachi, Tokyo, Japan) equipped with an energy dispersive spectrometer (EDS, Hitachi) was used to observe the microstructure of coatings and determine the content of elements, while the X-ray diffraction (XRD, Shimadzu 7000, Kyoto, Japan) was used to identify the constituent phase with Cu K$\alpha$ ($\lambda$ = 0.154 nm) radiation at a step of 0.02°. The microhardness was tested by a Vickers hardness tester (HVS-5, Lai Hua Ltd., Laizhou, China) with a load of 200 g and a duration time of 10 s. Sliding wear tests of the samples were performed on the universal tester by the ball-on-plate configuration under a dry condition, and the WC ball was chosen as the contact head. The friction and wear tests were carried out with a frequency of 5 Hz for 20 min.

## 3. Results and Discussion

Figure 1 shows the XRD results of FeCoCrNiAlTi$_x$ HEA coatings. It is obvious that both FCC and BCC$_1$ systems exist in the FeCoCrNiAl coating, which can be identified as (Fe, Ni) and Fe–Cr phases, respectively. However, the FCC system disappears in the FeCoCrNiAlTi$_{0.25}$ HEA coating and the AlNi, which is BCC forms. Ti is a BCC stabilizing element that restrains the formation of FCC. The AlNi is identified as BCC$_2$. Jiao et al. [13] gained a similar result in the experiment where AlCoCrFeNiTi$_x$ was deposited via arc melting and injection into a water-cooled copper mold.

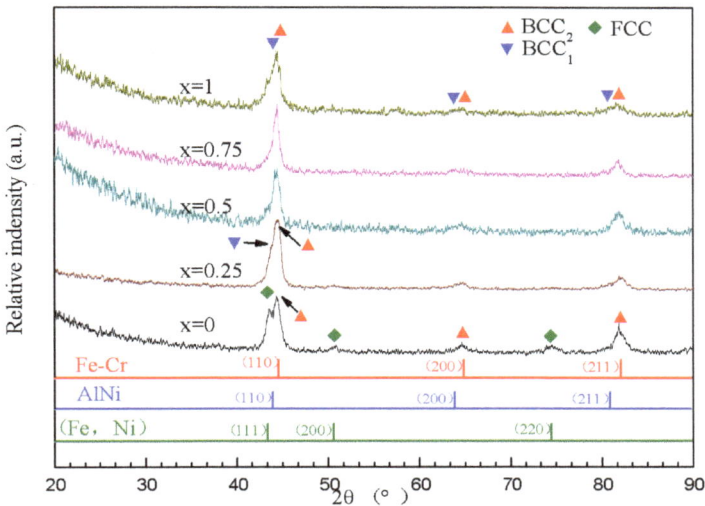

**Figure 1.** XRD results of FeCoCrNiAlTi$_x$ high-entropy alloy (HEA) coatings.

Figure 2 shows the cross-sectional SEM images of FeCoCrNiAlTi$_x$ HEA coatings. There are two different phases in the microstructure at $x = 0$. The chemical composition in different regions, as marked in Figure 2, is listed in Table 1. It can be seen by analyzing the EDS results in Table 1 that the components of Zone A and Zone B is very similar. Thus, Zone B is a simple hole. There are only eutectic FCC and BCC$_1$ structures here. The structure morphology changes significantly with the addition of Ti. At $x = 0.25$, the structure consists of the proeutectic phase $\alpha$ and the eutectic phase $\beta$. The proeutectic phase $\alpha$ exhibits an achrysanthemum-shaped microstructure, while the eutectic phase $\beta$ is a typical lamellar. According to the EDS results shown in Table 1, it can be concluded that the content of Ni and Al is high in phase $\alpha$ and the content of Fe and Cr is high in phase $\beta$. Combined with the former XRD results, it is obvious that the phase $\alpha$ is BCC$_1$, while the phase $\beta$ is the eutectic phase of BCC$_1$ and BCC$_2$, respectively. The segregation of Ni, Al, Fe, and Cr in different phases can be attributed to the mixing enthalpies of atomic pairs between these six elements, as it is known that the mixing enthalpies indicate a tendency to order or cluster. The mixing enthalpies represent atomic interactions between the solute elements and the base solvent. It is easy to occupy the cluster center for a solute because it shows a negative $\Delta H$ tendency and shows that a weak $\Delta H$ tends to take the glue site. Such a cluster-based formulism has been validated in many FCC and BCC alloys, which satisfies specific composition formulas in limited quantities [14].

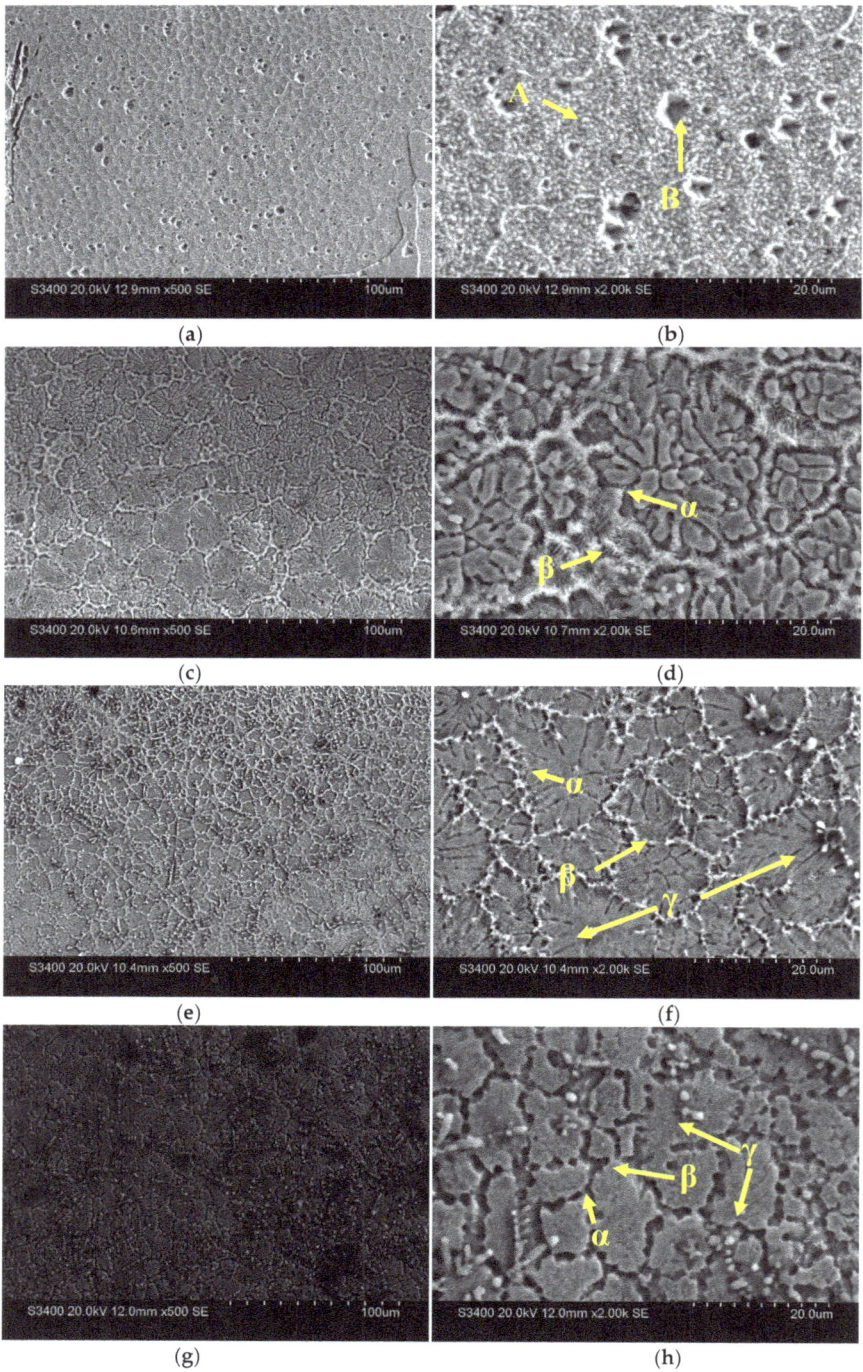

**Figure 2.** *Cont.*

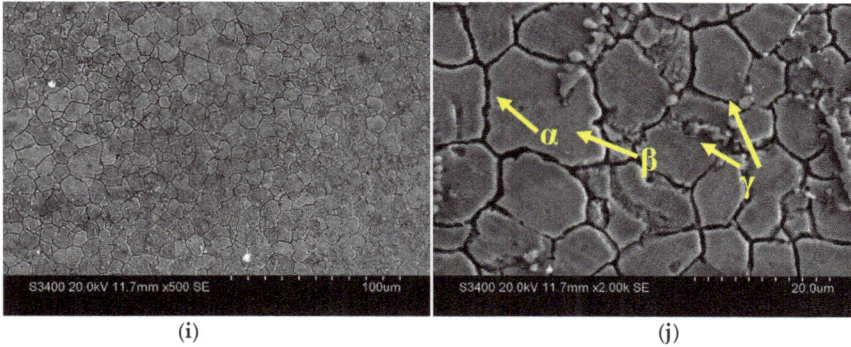

(i)                      (j)

**Figure 2.** Cross-sectional SEM images of FeCoCrNiAlTi$_x$ HEA coatings. (**a,b**) $x$ = 0; (**c,d**) $x$ = 0.25; (**e,f**) $x$ = 0.5; (**g,h**) $x$ = 0.75; (**i,j**) $x$ = 1.

**Table 1.** The chemical composition (at.%) in different regions, as marked in Figure 2.

| X | Zone | Fe | Co | Cr | Ni | Al | Ti |
|---|---|---|---|---|---|---|---|
| 0 | A | 32.26 | 24.84 | 10.76 | 23.54 | 8.60 | 0 |
| | B | 32.80 | 23.78 | 11.58 | 23.56 | 8.28 | 0 |
| 0.25 | $\alpha$ | 21.26 | 27.95 | 16.93 | 24.24 | 4.79 | 4.83 |
| | $\beta$ | 22.15 | 29.38 | 21.08 | 20.34 | 0.91 | 6.14 |
| 0.5 | $\alpha$ | 29.15 | 20.12 | 10.80 | 25.31 | 5.47 | 9.16 |
| | $\beta$ | 30.12 | 20.54 | 12.69 | 23.11 | 1.72 | 11.81 |
| | $\gamma$ | 15.55 | 13.48 | 10.23 | 14.31 | 2.12 | 44.31 |
| 0.75 | $\alpha$ | 33.75 | 18.03 | 12.58 | 19.71 | 4.13 | 11.83 |
| | $\beta$ | 39.73 | 19.31 | 14.30 | 11.21 | 2.74 | 12.72 |
| | $\gamma$ | 8.75 | 10.15 | 4.05 | 8.03 | 1.55 | 67.48 |
| 1 | $\alpha$ | 33.52 | 15.82 | 10.55 | 20.24 | 6.76 | 13.11 |
| | $\beta$ | 37.12 | 17.01 | 12.58 | 15.33 | 1.59 | 16.36 |
| | $\gamma$ | 11.05 | 6.64 | 5.42 | 0 | 2.67 | 74.22 |

The mixing enthalpy of atomic pairs between these six elements is listed in Table 2. As shown in Table 2, the mixing enthalpy of Al–Ni is −22 kJ/mol, so they have a good miscibility. However, the mixing enthalpies of Fe–Al and Cr–Al are −11 kJ/mol and −10 kJ/mol, respectively, which is relatively high among all of the pairs between Al and other elements. Therefore, Fe, Cr, and Al cannot have a good miscibility. Thus, Ni, Al and Fe, Cr segregate in different phases. According to the EDS results, the phase $\beta$ is a Ti-enriched phase, indicating that the solid solubility of Ti in phase $\beta$ is higher than that in phase $\alpha$. This result is in accordance with Shun et al. [15]. With the addition of Ti, the content of eutectic phase $\beta$ decreases during the increase of proeutectic phase $\alpha$. Thus, the eutectic point of BCC$_1$ and BCC$_2$ is between $x$ = 0 and $x$ = 0.25. As Ti content increases to $x$ = 0.5, some fine precipitates are precipitated out and the main component is Ti. The precipitates can hinder the dislocation glide, thereby enhancing the hardness and the strength. At $x$ = 1, equiaxed and fine grained microstructure is obtained in the structure and the eutectic phase $\beta$ almost disappears.

**Table 2.** The mixing enthalpy of atomic pairs between these six elements (kJ/mol) [16].

| Element | Fe | Co | Cr | Ni | Al | Ti |
|---|---|---|---|---|---|---|
| Fe | 0 | – | – | – | – | – |
| Co | −1 | 0 | – | – | – | – |
| Cr | −1 | −4 | 0 | – | – | – |
| Ni | −2 | 0 | −7 | 0 | – | – |
| Al | −11 | −19 | −10 | −22 | 0 | – |
| Ti | −17 | −28 | −7 | −35 | −30 | 0 |

Although there are several principle elements, the phase composition of the HEAs is relatively simple. There are only a few kinds of solid solutions in the coatings. The phase number of $FeCoCrNiAlTi_x$ HEA coatings in this paper is much less than that calculated by the Gibbs phase rule. This phenomenon can be attributed to the high entropy effect, which can restrain the formation of complex intermetallic compounds. Otto et al. [17] and Yao et al. [18] also found this phenomenon. According to the Gibbs free energy formula:

$$\Delta G_{mix} = \Delta H_{mix} - T\Delta S_{mix} \tag{1}$$

where $\Delta H_{mix}$ is the mixing enthalpy, $\Delta S_{mix}$ is the mixing entropy, and $T$ is the absolute temperature. It can be seen that the $\Delta G_{mix}$ will decrease if the $\Delta S_{mix}$ increases. The high $\Delta G_{mix}$ is the driving force for the formation of intermetallic compounds. Thus, the high entropy effect can decrease the $\Delta G_{mix}$ of the system and contribute to the formation of saturated or supersaturated solid solution, thereby enhancing the solution strength effect and the mechanical properties.

Figure 3 shows the results of the microhardness test along the cross section of the $FeCoCrNiAlTi_x$ HEA coatings. The average microhardness of the HEA coatings at $x = 0$, 0.25, 0.5, 0.75, and 1 are 432.73 HV, 548.81 HV, 651.03 HV, 769.20 HV, and 966.29 HV, respectively. It is obvious that the average hardness of $FeCoCrNiAlTi_x$ HEA coatings increases as the content of Ti increases. The average hardness reaches its maximum point for the HEA at $x = 1$, which is more than twice that of the HEA at $x = 0$. The content of Ti in $BCC_1$ and $BCC_2$ persistently increases (as shown in Figure 1) with the continuous addition of Ti. The atomic radii of Fe, Co, Ni, Cr, Al, and Ti are 124 pm, 125 pm, 121 pm, 124 pm, 143 pm, and 145 pm, respectively. It was easily observed that the radius of Ti is the highest among all elements. With the addition of Ti, the solid solute effect increases, leading to serious lattice distortion. Thus, the solid solution strengthening effect of Ti may be the most important reason, which contributes to the increase in hardness. In addition, the second phase strengthening effect caused by the separation of Ti is also an important reason. It can be seen in Figure 3 that there is a softening region near the fusion line between the coating and the substrate. In this zone, the grains are coarse, which affect the mechanical properties of the coatings.

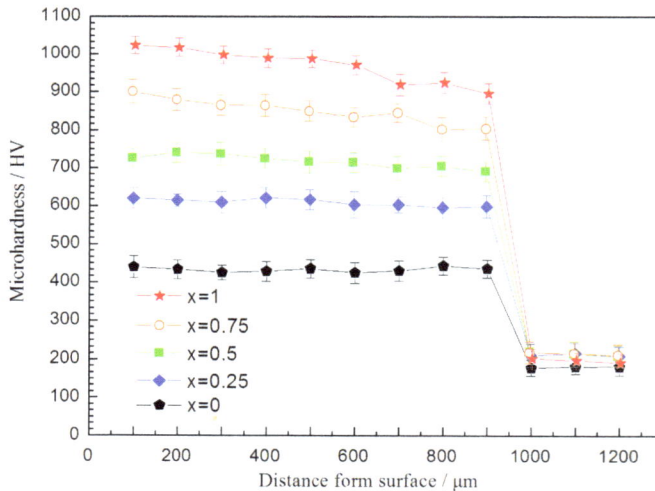

**Figure 3.** Results of the microhardness test along the cross section of the $FeCoCrNiAlTi_x$ HEA coatings.

Figure 4 shows the worn surface morphologies of FeCoCrNiAlTi$_x$ HEA coatings at the microscopic level. The surface tends to be smooth with the addition of Ti. Typically, the abrasion rate is connected with the resistance force at the contact surface during wear [19]. A higher resistance force usually leads to a higher abrasion rate. Figure 5 shows the variations in the friction coefficient curves of FeCoCrNiAlTi$_x$ coating as a function of wear test time. The mass loss is proportional to the friction coefficient in Figure 6. According to the surface morphology, there are no obvious pits or other morphology features of adhesive wear. There are grooves and debris on the worn surface. Thus, the worn mechanism is the abrasive wear. At $x = 0$, the grooves are deep, and there is substantial debris. The mass loss is also higher for the HEA at $x = 0$, so the wear resistance is lower compared to the HEAs with higher Ti concentrations. At $x = 0.25$, the amount of debris decreases, and the grooves are shallow. It is obvious that the wear resistance of the coating is enhanced with the addition of Ti. At $x = 0.5$ and 0.75, the deep grooves disappear. At $x = 1$, only a few shallow grooves can be seen, and there is no debris. The mass loss of FeCoCrNiAlTi$_x$ coatings is the lowest at $x = 1$, according to Figure 6, indicating optimum wear resistance.

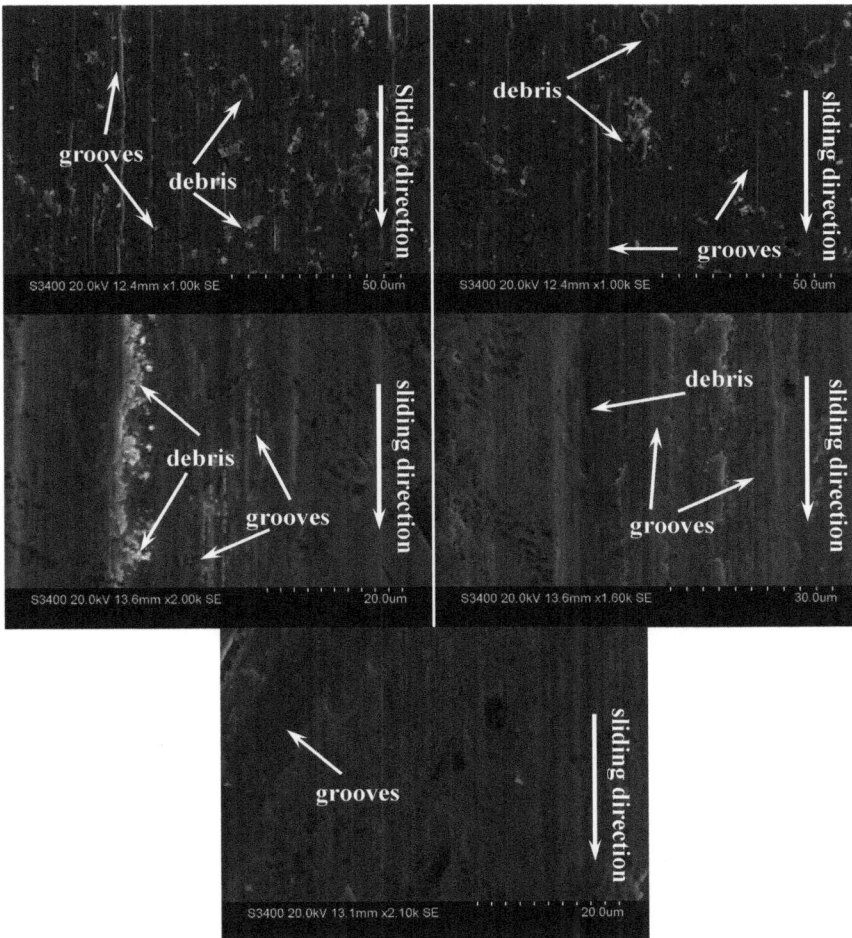

**Figure 4.** The morphology of FeCoCrNiAlTi$_x$ HEA coatings after wear.

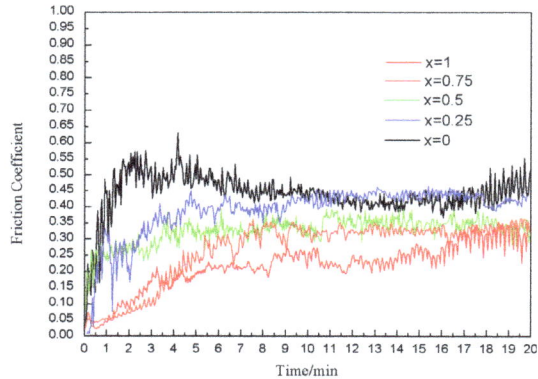

**Figure 5.** Friction coefficient of coating.

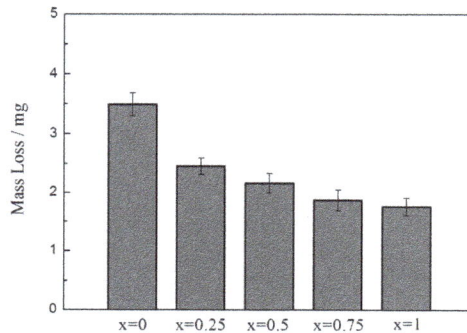

**Figure 6.** Mass loss of FeCoCrNiAlTi$_x$ coatings.

## 4. Conclusions

After studying the effect of the content of Ti on the performance of FeCoCrNiAlTi$_x$, three conclusions can be drawn as follows:

- The structure of FeCoCrNiAlTi$_x$ coatings consists of two kinds of BCC phases. With the continuous increase in Ti content, the number of eutectic structure decreases, accompanying the increase in the proeutectic phase and the Ti-rich second phase is precipitated out. The eutectic point of BCC$_1$ and BCC$_2$ is between $x = 0$ and $x = 0.25$.
- The solution strength effect and the second phase strengthening effect on FeCoCrNiAlTi$_x$ coatings are enhanced as Ti content increases. The average microhardness is also increased.
- The friction coefficient of FeCoCrNiAlTi$_x$ coatings decreases as Ti content increases, while the wear resistance increases. The HEA coatings at $x = 1$ exhibit the best wear resistance among the tested samples.

**Acknowledgments:** This work was financially supported by the Natural Science Foundation of Liaoning Province (No. 201602553), the Chinese National Natural Science Foundation (No. 51301112, No.51401129), and the China Postdoctoral Science Foundation (2015M 571327).

**Author Contributions:** Bin He performed the experiments and wrote the manuscript; Nannan Zhang designed the experimental process; Danyang Lin analyzed the data; Yue Zhang assisted in SEM analysis; Fuyu Dong assisted in paper review; Deyuan Li guided the direction of the research.

**Conflicts of Interest:** The authors declare no conflict of interest.

## References

1. Shen, W.J.; Tsai, M.H.; Yeh, J.W. Machining performance of sputter-deposited $(Al_{0.34}Cr_{0.22}Nb_{0.11}Si_{0.11}Ti_{0.22})_{50}N_{50}$ high-entropy nitride coatings. *Coatings* **2015**, *5*, 312–325. [CrossRef]
2. Zhang, Y.; Zuo, T.T.; Tang, Z.; Gao, C.; Dahmen, K.A.; Liaw, P.K.; Lu, Z.P. Microstructures and properties of high-entropy alloys. *Prog. Mater. Sci.* **2014**, *61*, 1–93. [CrossRef]
3. Gludovatz, B.; Hohenwarter, A.; Catoor, D.; Chang, E.H.; George, E.P.; Ritchie, R.O. A fracture-resistant high-entropy alloy for cryogenic applications. *Science* **2014**, *345*, 1153–1158. [CrossRef] [PubMed]
4. He, Y.Z.; Zhang, J.L.; Zhang, H.; Song, G.S. Effects of different levels of boron on microstructure and hardness of $CoCrFeNiAl_xCu_{0.7}Si_{0.1}$ by high-entropy alloy coatings by laser cladding. *Coatings* **2017**, *7*, 7. [CrossRef]
5. Tsai, M.H.; Yeh, J.W. High-entropy alloys: A critical review. *Mater. Res. Lett.* **2014**, *2*, 107–123. [CrossRef]
6. Yeh, J.W. Physical Metallurgy of High-Entropy Alloys. *JOM* **2015**, *67*, 2254–2261. [CrossRef]
7. Duan, H.; Wu, Y.; Hua, M.; Yuan, C.Q.; Wang, D.; Tu, J.S.; Kou, H.C.; Li, J. Tribological properties of AlCoCrFeNiCu high-entropy alloy in hydrogen peroxide solution and in oil lubricant. *Wear* **2013**, *297*, 1045–1051. [CrossRef]
8. Tian, F.Y.; Varga, L.K.; Chen, N.X.; Shen, J.; Vitos, L. Empirical design of single phase high-entropy alloys with high hardness. *Intermetallics* **2015**, *58*, 1–6. [CrossRef]
9. Chen, X.Y.; Yan, L.; Karnati, S.; Zhang, Y.L.; Liou, F. Fabrication and characterization of $Al_xCoFeNiCu_{1-x}$ high entropy alloys by laser metal deposition. *Coatings* **2017**, *7*, 47. [CrossRef]
10. An, X.; Liu, Q.; Zheng, B. Microstructure and properties of laser cladding high entropy alloy $MoFeCrTiWAl_xSi_y$ coating. *Infra. Laser Eng.* **2014**, *43*, 1140–1144.
11. Zhang, C.; Wu, G.F.; Dai, P.Q. Phase transformation and aging behavior of $Al_{0.5}CoCrFeNiSi_{0.2}$ high-entropy alloy. *J. Mater. Eng. Perform.* **2015**, *24*, 1918–1925. [CrossRef]
12. Tang, Z.; Gao, M.C.; Diao, H.Y.; Yang, T.F.; Liu, J.P.; Zuo, T.T; Zhang, Y.; Lu, Z.P.; Cheng, Y.Q.; Zhang, Y.W.; et al. Aluminum alloying effects on lattice types, microstructures, and mechanical behavior of high-entropy alloys system. *JOM* **2013**, *65*, 1848–1858. [CrossRef]
13. Jiao, Z.M.; Ma, S.G.; Chu, M.Y.; Yang, H.J.; Wang, Z.H; Zhang, Y.; Qiao, J.W. Superior mechanical properties of $AlCoCrFeNiTi_x$ high-entropy alloys upon dynamic loading. *J. Mater. Eng. Perform.* **2016**, *25*, 451–456. [CrossRef]
14. Wang, Q.; Ma, Y.; Jiang, B.B.; Li, X.N.; Shi, Y.; Dong, C.; Liaw, P.K. A cuboidal B2 nanoprecipitation-enhanced body-centered-cubic alloy $Al_{0.7}CoCrFe_2Ni$ with prominent tensile properties. *Scripta Mater.* **2016**, *120*, 85–89. [CrossRef]
15. Shun, T.T.; Hung, C.H.; Lee, C.F. The effects of secondary elemental Mo or Ti addition in $Al_{0.3}CoCrFeNi$ high-entropy alloy on age hardening at 700 °C. *J. Alloys Compd.* **2010**, *495*, 55–58. [CrossRef]
16. Takeuchi, A.; Inoue, A. Classification of bulk metallic glasses by atomic size difference, heat of mixing and period of constituent elements and its application to characterization of the main alloying element. *Mater. Trans.* **2005**, *46*, 2817–2829. [CrossRef]
17. Otto, F.; Dlouhý, A.; Pradeep, K.G; Kuběnová, M.; George, E.P. Decomposition of the single-phase high-entropy alloy CrMnFeCoNi after prolonged anneals at intermediate temperatures. *Acta Mater.* **2016**, *112*, 40–52. [CrossRef]
18. Yao, M.J.; Pradeep, K.G.; Tasan, C.C.; Raabe, D. A novel, single phase, non-equiatomic FeMnNiCoCr high-entropy alloy with exceptional phase stability and tensile ductility. *Scripta Mater.* **2014**, *72–73*, 5–8. [CrossRef]
19. Yu, Y.; Liu, W.; Zhang, T.; Li, J.; Wang, J.; Kou, H.; Li, J. Microstructure and tribological properties of $AlCoCrFeNiTi_{0.5}$, high-entropy alloy in hydrogen peroxide solution. *Metall. Mater. Trans. A* **2014**, *45*, 201–207. [CrossRef]

# coatings

MDPI

*Article*

# Effect of Coating Palm Oil Clinker Aggregate on the Engineering Properties of Normal Grade Concrete

**Fuad Abutaha, Hashim Abdul Razak *and Hussein Adebayo Ibrahim**

StrucHMRS Group, Department of Civil Engineering, Faculty of Engineering, University of Malaya, Kuala Lumpur 50603, Malaysia; abutaha.fuad@siswa.um.edu.my (F.A.); adebayor@siswa.um.edu.my (H.A.I.)
* Correspondence: hashim@um.edu.my; Tel.: +60-3-7967-5233; Fax: +60-3-7967-5318

Academic Editor: Alicia Esther Ares
Received: 10 August 2017; Accepted: 5 September 2017; Published: 21 October 2017

**Abstract:** Palm oil clinker (POC) is a waste material generated in large quantities from the palm oil industry. POC, when crushed, possesses the potential to serve as an aggregate for concrete production. Experimental investigation on the engineering properties of concrete incorporating POC as aggregate and filler material was carried out in this study. POC was partially and fully used to replace natural coarse aggregate. The volumetric replacements used were 0, 20%, 40%, 60%, 80%, and 100%. POC, being highly porous, negatively affected the fresh and hardened concrete properties. Therefore, the particle-packing (PP) method was adopted to measure the surface and inner voids of POC coarse aggregate in the mixtures at different substitution levels. In order to enhance the engineering properties of the POC concrete, palm oil clinker powder (POCP) was used as a filler material to fill up and coat the surface voids of POC coarse, while the rest of the mix constituents were left as the same. Fresh and hardened properties of the POC concrete with and without coating were determined, and the results were compared with the control concrete. The results revealed that coating the surface voids of POC coarse with POCP significantly improved the engineering properties as well as the durability performance of the POC concrete. Furthermore, using POC as an aggregate and filler material may reduce the continuous exploitation of aggregates from primary sources. Also, this approach offers an environmental friendly solution to the ongoing waste problems associated with palm oil waste material.

**Keywords:** coating; lightweight aggregate; palm oil clinker (POC); palm oil clinker powder (POCP); waste material

## 1. Introduction

The construction industry has been identified to be one of the largest industries worldwide. Currently, an enormous worldwide development is occurring in this industry, especially in developing countries as a result of rapid industrial and economic developments leading to the improved standard of living and infrastructural development [1]. Nowadays, the scarcity of natural resources and the rising costs of raw materials have induced researchers to focus more on utilizing solid wastes and by-products as raw material in concrete production [1,2]. Economic and environmental benefits are some of the factors that determine the viability of using solid waste. From an economic standpoint, using solid waste is cheaper as compared to the costs of using natural resources or even the costs of producing new material [3]. Consequently, natural resources can be preserved and there will be a significant reduction in waste being discharged to the environment [4]. The solid wastes and by-products, when properly used, have shown to be a comparable construction raw material [4]. Palm oil clinker (POC) is a waste material generated in large quantities from the palm oil industry. POC, when crushed, has the potential to serve as a lightweight aggregate for concrete production.

## 2. Pam Oil Clinker (POC)

Malaysia is one of the primary producers of palm oil, contributing more than half of the world's palm oil on a yearly basis [5]. In 2012, it was reported that the total production of crude palm oil was more than 18.7 million tons [6]. The extraction of useful material from these plants generates various types and forms of waste material, which need to be disposed of appropriately. Generally, they comprise of ash, grains, wastewater, and shells in large combined chunks [7]. POC is a palm oil shell incineration by-product in the form of a lightweight material (Figure 1). The large chunk of POC is flaky, irregularly shaped, and porous with a rough and sharp broken surface, as seen in Figure 2. POC, when crushed, represents a lightweight aggregate that can be potentially used in concrete production [8]. Kanadasan et al. (2015) [9] reported that the similarity of the particle size distribution and the grading features of sand and POC fine aggregate indicate the suitability of POC fine as a suitable substitution for sand in concrete production. Abdullahi et al. (2008) [10] prepared trial mix proportions for POC, and showed that it is possible to use POC as an aggregate in the mix design of concrete without adding admixture. Kanadasan et al. (2014) [7] reported that POC aggregates, which are often seen as a waste material for landfilling, performed satisfactorily as an aggregate material in the production of self-compacting concrete. Also, a study conducted by Ibrahim et al. (2016) [11] indicated the feasibility of incorporating POC aggregate as a suitable replacement for natural aggregates in Pervious Concrete (PC) production. In this study, palm oil clinker powder (POCP) was used as a filler material to fill up and coat the surface voids of POC coarse so as to enhance the engineering properties of normal grade POC concrete. The aim was to examine the applicability of palm oil waste as a natural aggregate replacement in concrete production towards improving the sustainability of the construction industry.

**Figure 1.** Solid waste of palm oil mill.

**Figure 2.** Raw POC collected from palm oil mill.

## 3. Materials

### 3.1. Aggregates

Three types of aggregate were used in this study, which includes POC, river sand as natural fine aggregate, and crushed granite rocks as a natural coarse aggregate. POC was collected from a local palm oil mill in the form of a large chunk. It was then crushed using a Jaw crusher machine and sieved to be used as a replacement for natural coarse aggregate (Figure 3). The physical properties of all aggregates used in this study are tabulated in Table 1.

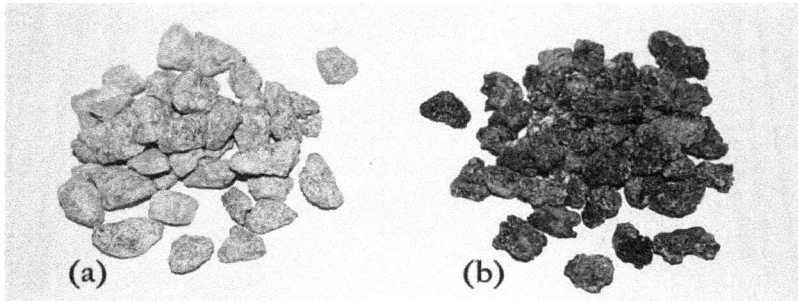

**Figure 3.** Coarse aggregate: (**a**) Granite; (**b**) Palm oil clinker (POC).

**Table 1.** Physical characteristics of the aggregates.

| Properties | Aggregates | | |
|---|---|---|---|
| | River Sand | Coarse Aggregates | |
| | | Granite | POC |
| Aggregate size (mm) | <4.75 | 4.75–14 | 4.75–14 |
| Specific gravity | 2.66 | 2.65 | 1.73 |
| Water absorption (%) | 0.39 | 0.58 | $3 \pm 2$ |
| Moisture content (%) | 0.08 | 0.28 | $1 \pm 0.5$ |
| Aggregate crushing Value (%) | – | 17.93 | 56.44 |
| Aggregate crushing value (Ten per cent fines) | – | – | 16.99 |
| Bulk Density (kg/m$^3$) | 1301 | 1294 | 732 |

### 3.2. Powders

Ordinary Portland Cement (OPC), equivalent to ASTM Type I, was used as the main binding material. POC powder was prepared by grinding POC into a fine powder form. POCP can be assumed to have a similar fineness with cement [12]. The particle size distribution curves of POCP and OPC are comparable and most of the particles sizes are less than 100 μm [13]. In this study, POCP was used as filler material to fill up and coat the surface voids of POC particles. A comparison of the physical properties and the chemical composition of POCP and OPC are presented in Table 2.

### 3.3. Admixtures

To enhance the concrete workability, Sika ViscoCrete 2199 from Sika Kimia Sdn Bhd, Kuala Lumpur, Malaysia, which is a modified polycarboxylate type was used as a high range water reducing admixture in the study. This admixture is chloride free according to BS 5075 and is compatible with all types of Portland cement.

**Table 2.** Chemical composition and physical properties of Ordinary Portland Cement (OPC) and palm oil clinker powder (POCP) [12].

| Properties | OPC | POCP |
|---|---|---|
| *Chemical Composition (%)* | | |
| CaO | 64 | 6.37 |
| SiO$_2$ | 20.29 | 59.9 |
| SO$_3$ | 2.61 | 0.39 |
| Fe$_2$O$_3$ | 2.94 | 6.93 |
| Al$_2$O$_3$ | 5.37 | 3.89 |
| MgO | 3.13 | 3.3 |
| P$_2$O$_5$ | 0.07 | 3.47 |
| K$_2$O | 0.17 | 15.1 |
| TiO$_2$ | 0.12 | 0.29 |
| Mn$_2$O$_3$ | 0.12 | – |
| Na$_2$O | 0.24 | – |
| others | 0.94 | 0.36 |
| Loss on ignition | 1.4 | 1.89 |
| *Physical Properties* | | |
| Specific gravity (g/cm$^3$) | 3.15 | 2.59 |
| Particle Distribution | – | – |
| Average size, D (V, 0.5) | 27.98 µm | 20.97 µm |
| Passing 10.48 µm | 27.58 | 37.86 |
| Retained 10.48 µm, Passing 48.27 µm (%) | 45.80 | 34.05 |
| Retained 48.27 µm (%) | 26.62 | 28.09 |

## 4. Experimental Program

The experimental program involved determining the engineering properties of the concrete made with POC as a partial and full replacement of natural coarse aggregate with and without the surface coating. POC was collected in the form of large chunks from the palm oil mill factory, as shown in Figure 4. It was then crushed using a Jaw crusher machine, and it was sieved to the required size. The mix design was based on the Department of Environment (DOE) method to produce concrete grade 40 with slump in the range of $100 \pm 25$ mm. The volumetric replacement of granite coarse aggregate with POC adopted in this study are 0, 20%, 40%, 60%, 80%, and 100%. All of the mixes had a constant cement content and water to cement ratio of 420 kg/m$^3$ and 0.53, respectively, so as to observe only the effect of POC incorporation on the fresh and hardened concrete properties. Details of the constituent materials proportion at different substitution levels are presented in Table 3.

**Figure 4.** Palm oil clinker (POC).

Table 3. Mixture proportion for different replacement level of coarse aggregate.

| Replacement Level | ID | w/c Ratio | Mix Proportion (kg/m³) | | | | |
|---|---|---|---|---|---|---|---|
| | | | OPC | Fine Aggregate | | Coarse Aggregate | |
| | | | | River Sand | POC | Granite | POC |
| Control Mix | M0 | 0.53 | 420 | 760 | – | 1007 | – |
| Coarse Aggregate Replacement (Series POC) | | | | | | | |
| 20% | POC20 | 0.53 | 420 | 760 | – | 806 | 131 |
| 40% | POC40 | 0.53 | 420 | 760 | – | 604 | 263 |
| 60% | POC60 | 0.53 | 420 | 760 | – | 402 | 394 |
| 80% | POC80 | 0.53 | 420 | 760 | – | 201 | 526 |
| 100% | POC100 | 0.53 | 420 | 760 | – | – | 657 |

POC, being highly porous, and due to no standard guidelines available in literature on how to choose the quantity of powder required to coat the surface voids of such an aggregate, the particle packing (PP) method was adopted to determine the volume of voids due to the porosity of POC at different substitution levels. The PP method gives an assessment of the additional voids when POC is substituted with the natural aggregate. The steps for using the PP method to measure the voids are as follows:

- Phase 1: All the aggregate particles were checked to ensure they have been soaked in water for 24 h. They were later brought into the saturated surface dry (SSD) condition to avoid any loss of fluid through absorption during the PP test.
- Phase 2: The combination of the aggregates, i.e., POC coarse, granite coarse and river sand with proportion based on the mix designed by the DOE method, as tabulated in Table 3, was prepared on the baseplate. They were thoroughly mixed by using a scoop and trowel to get a homogenous mix. It was later placed into a container in a loosely packed state, as shown in Figure 5.
- Phase 3: A known volume of clean water was prepared and it was subsequently poured slowly into the container filled with the aggregates.
- Phase 4: Once the water level reached the top surface of the container, the water level is checked consecutively every 30 s for a period of 2 min. This is basically to allow for water to fill up all the voids between the aggregates. Water is constantly added if there is a reduction in the level. The amount of water utilized represents the total amount of voids present.

Figure 5. Schematic diagram of PP test.

*Determination of the Required POCP*

To determine the particle packing density of concrete, small particles should be selected to fill up the voids between the large particles [14]. In this study, palm oil clinker powder (POCP), which was obtained by a ball mill grinding process of POC, was selected as the suitable filler material to coat the surface voids of POC coarse in order to enhance the fresh and hardened properties of the POC concrete. It is important to design concrete structures and mixtures in such a way that any negative environmental impact is minimized. Thus, using POCP serves as an environmentally friendly alternative and is also a means of maximizing the use of palm oil waste in the concrete. The general procedure for determining the POCP required for each substitution level of POC coarse is shown in Figure 6, and the detailed calculations are given in Appendix A.

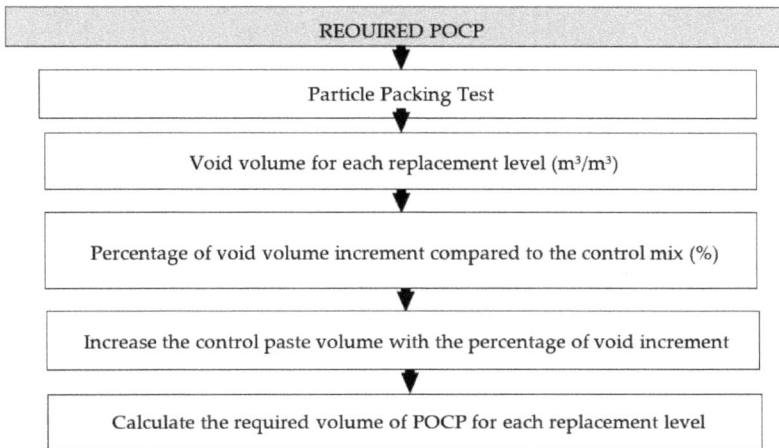

Figure 6. Flow chart to determine the required POCP.

Based on the PP results, the final mix proportions after incorporating POCP to the six levels of POC concrete is presented in Table 4. Figure 7 shows the total powder and superplasticizer (SP) dosage required for different substitution levels of POC to obtain a constant slump range of $100 \pm 25$ mm. A comparison was conducted to determine the effect of the filling-ability of POCP on fresh and hardened properties of POC concrete.

**Table 4.** Final mix proportion of POCP concrete.

| Replacement Level | ID | Cement (kg/m³) | POCP (kg/m³) | w/p Ratio | Fine Aggregate (kg/m³) | Coarse Aggregate (kg/m³) | | |
|---|---|---|---|---|---|---|---|---|
| | | | | | River Sand | Granite | POC | |
| Control Mix | M0 | 420 | 0 | 0.53 | 760 | 1007 | 0 | |
| 20% | POCP20 | 420 | 70 | 0.51 | 760 | 806 | 131 | |
| 40% | POCP40 | 420 | 93 | 0.48 | 760 | 604 | 263 | |
| 60% | POCP60 | 420 | 108 | 0.46 | 760 | 402 | 394 | |
| 80% | POCP80 | 420 | 156 | 0.45 | 760 | 201 | 526 | |
| 100% | POCP100 | 420 | 203 | 0.43 | 760 | 0 | 657 | |

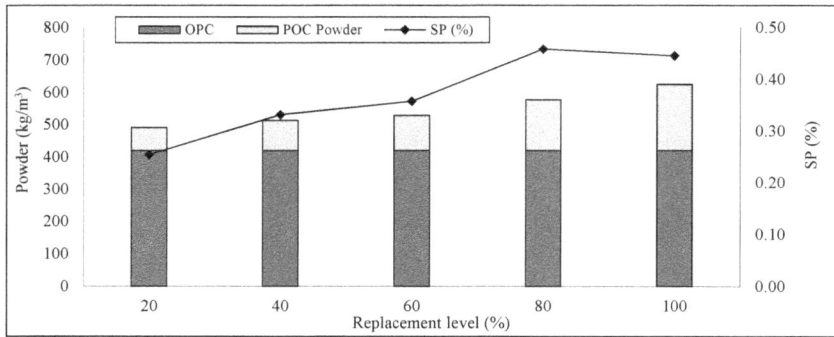

Figure 7. POCP and SP dosage for each replacement level.

## 5. Coating Process and Concrete Mixing

The volume of POCP required for each replacement level of POC coarse were calculated and presented in Table 4. POCP was calculated by determining the volume of the additional voids with reference to granite. Taking into consideration the substantial volume of voids and the irregular shape of POC coarse aggregate. The combination of POC and POCP with a proportion based on the PP mix design, as tabulated in Table 4, was prepared. They were thoroughly dry mixed for 4–5 min using a pan type mixer in order to fill up the voids and properly coat the POC particles, as shown in Figure 8.

(a)                                                      (b)

10mm

Figure 8. POC particle: (a) Pre-coating; (b) After coating.

The remaining aggregates i.e., sand and granite with cement, were added and allowed to dry mix for an additional 2 min. After that, two third of the mixing water was added and allowed to mix for 3 min. The remaining water and SP were gradually added to the mixture. The concrete mix was subjected to additional mixing for about 5 min to ensure a homogenous mix was obtained. Subsequently, fresh properties were determined by measuring the concrete workability and fresh density after completing the mixing process. The quantity of concrete was always prepared 20% in excess of the required amount. The fresh concrete was casted in steel moulds using shovels. Prior to casting, the surfaces of the moulds were cleaned and a thin layer of oil was applied to the interior faces of the moulds to facilitate the de-moulding process. Fresh concrete was casted in three layers and each layer was compacted using a vibration table. After the final layer had been compacted, the top was levelled to provide a smooth and flat surface, and it was then covered with gunnysacks to prevent moisture loss and minimize the plastic shrinkage. The specimens were de-moulded after 24 h and then subjected to full water curing until the date of the hardened concrete test. The final value of the mechanical properties was determined by taking the average of three identical specimens and the shrinkage value for each age is the average value of nine readings.

## 6. Results and Discussion

### 6.1. Workability

The consistency of the concrete was assessed by the measure of slump in this study. Slump results of the POC and POCP concrete mixes are depicted in Figure 9. The workability of the mixes was affected by incorporating the POC coarse. Increasing the substitution ratio of POC decreased the workability of the mix. The mixes up to a 40% replacement of POC achieved the target slump range of 100 ± 25 mm. Meanwhile, the mixes with more than 40% of POC coarse were found to be less cohesive, having a high segregation and being somewhat harsher than the corresponding conventional concrete. The reduction of workability can be attributed to the particle shape and rough surface, as well as the sharp broken edges of POC. The irregular shape of POC resulted in a higher surface area increasing the demand for extra paste volume to ensure a good workability and to avoid segregation. Based on the experimental study by Koehler and Fowler (2007) [15], they concluded that the workability of a mix is a function of the aggregate characteristics, the paste volume, and the rheology of the paste. However, it is obvious that the workability of the POCP concrete was improved when the addition of POCP was used as a filler material to fill up the voids of POC particles. The observed improvement in workability can be partly attributed to the higher paste volume that makes the concrete cohesive enough to be handled without segregation or bleeding. The POCP helped to coat the POC particles, filling the gaps between the aggregates, thereby providing a better chance for aggregate lubrication. It can be seen that the mixes incorporating a higher content of POC coarse tend to require a higher POCP content to give extra lubrication to the POC aggregate, as well as higher dosages of SP to make the mixes more cohesive and to achieve target slump range of 100 ± 25 mm.

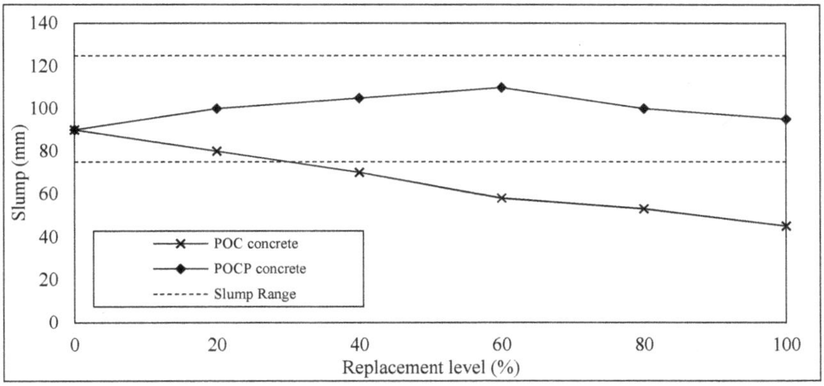

**Figure 9.** Slump values of POC and POCP concrete mixes.

### 6.2. Fresh Density

The fresh density of the POC and POCP concrete, respectively, ranged from 2032 to 2293 kg/m$^3$, as shown in Figure 10. The unit weight of the concrete with POC aggregates is inversely proportional to the replacement level. Increasing the amount of POC in the mix reduced the unit weight of the concrete. The maximum reduction of fresh density was at full replacement, which registered a value of 14% less than that of normal concrete i.e., 2379 kg/m$^3$. The low bulk density of the POC coarse resulted in the reduction of the unit weight of the POC concrete. The existence of a large number of voids and pores contributed significantly to the light nature of the POC aggregate. Kanadasan and

Razak [16] reported that the lower unit weight, coupled with the porous nature of the POC aggregate, directly resulted in a lower mass per volume of POC Self Compacting Concrete (SCC) Their results revealed that a full replacement of POC produced a concrete with a density of less than 2000 kg/m$^3$, which is approximately 16% lower than the control mix.

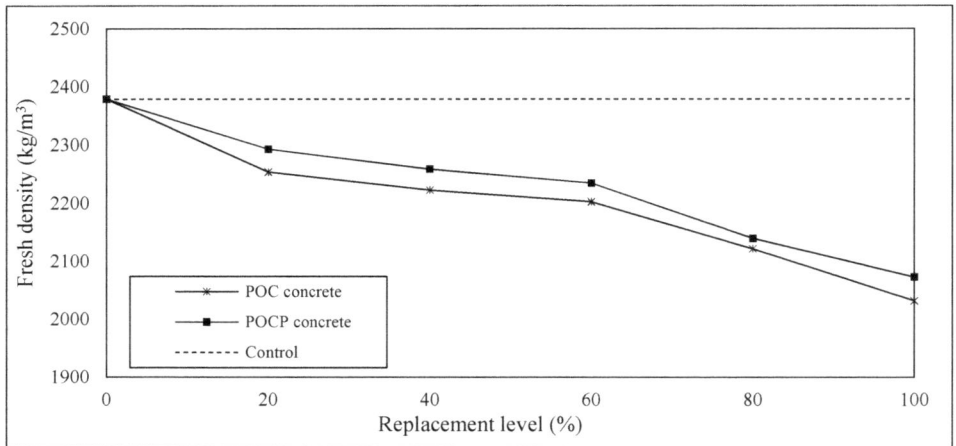

**Figure 10.** Fresh density of POCP concrete and POC concrete.

## 6.3. Compressive Strength

The compressive strength of the concrete was determined in accordance with BS EN 12390-3. In general, there was reduction in compressive strength when the POC coarse aggregate was used. The concrete strength was lower when the volume of the POC was higher than the volume of the conventional aggregate. At 28 days, the compressive strength of the POC concrete ranged between 33 and 39.45 MPa. The maximum reduction in strength was at full replacement of POC, i.e., approximately 30% lost with respect to the control concrete. The strength and stiffness of POC was much lower than the normal coarse aggregate due to its porous property, which significantly affected the concrete's strength carrying capacity. This is also attributed to the fact that less matrix is available to fill the pores within the POC aggregates, leading to a higher total porosity in the concrete as observed in Figure 11. Abutaha et al. (2016) [17], and Ibrahim et al. (2017) [18] reported that the porosity of POC and the lower aggregate crushing value negatively affected the compressive strength when compared to the control mix. This also corroborates the results of the study by Rashid et al. (2012) [1] on the utilization of LWA from waste material by using crushed clay bricks as a coarse aggregate replacement. The study reported that the replacement resulted in a strength reduction of 9.6% and 32.9% at 50% and 100% replacement levels, respectively. Furthermore, lightweight concrete incorporating aggregate made from sewage sludge waste showed good hardened properties whereby they were able to produce 73% of the strength when compared to that of the control concrete [19]. However, it can be observed in this study that a significant reduction in the compressive strength was avoided with the use of POCP as a filler material to fill the voids of POC coarse. Additionally, POCP significantly improved the compressive strength of the POC concrete by providing a sufficient amount of paste to coat the POC surface voids.

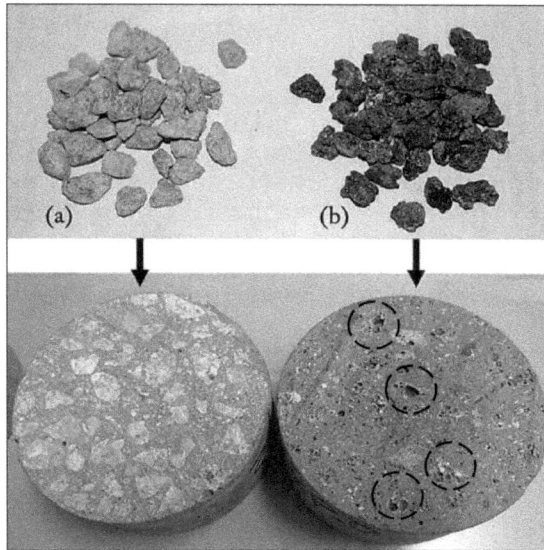

**Figure 11.** Difference in porosity of concrete. (**a**) Granite; (**b**) POC.

The influence of the additional POCP on the compressive strength of the POC concrete mixes is plotted in Figure 12. At 28 days, the POCP concrete had compressive strength values ranging between 40.52 and 51.27 MPa with maximum strength obtained at 20% POC coarse replacement. Compared to the mixes without POCP (pre-coating), the strength increased by 20% to 30%. Figure 13 illustrates the development of the POCP concrete strength. Figure 14 shows the variation of relative strength, which can be defined as the ratio of the tested specimen strength to that of control at the same age. For all replacement levels, the maximum relative strength occurred on the 3rd day for the mix of POCP20, and decreased until it reached a relatively constant value at 56 days and beyond. At a 20% replacement of POC coarse, the compressive strength improved due to the lesser percentage of POC in the mix when compared to the other replacement levels. Therefore, the effect of ACV of POC on the compressive strength of the concrete was not pronounced. The combined effect of POCP and the low percentage replacement of POC coarse ensured that POCP20 exhibited superior properties as compared to other replacement levels. However, at the age of 28 days, the mixes with up to an 80% replacement ratio of POC coarse exceeded the control strength. Meanwhile the mix at full replacement achieved 90% of the control strength, which continued to increase with age until it became close to the control strength value after 90 days, as shown in Figure 14. These results indicate that the inclusion of POCP gave a tremendous contribution to the strength property of the POC concrete. Thus, the finer POCP particles have better voids-filling ability, resulting in low void space that leads to a higher compressive strength. Furthermore, the use of superplasticizer also increases the strength by lowering the quantity of mixing water and increasing the flowing ability. It also contributed to achieving denser packing and a lower porosity of the concrete, and thus assisted in producing a higher strength and good durable concrete.

*6.4. Splitting Tensile Strength*

A splitting tensile strength test was carried out in accordance with BS 1881: Part 117 on a cylinder of (100 mm diameter × 200 mm length) at the age of 7, 28, 56, and 90 days. Splitting tensile strength results of the POC concrete generally showed a trend similar to that observed in the compressive strength. The replacement of POC with natural coarse aggregate led to a reduction in the splitting tensile property, as shown in Figure 15. The higher the contents of the POC coarse, the lower the splitting tensile value. At 28 days, the splitting tensile of the POC concrete was in the range of 2.61 to 3.28 MPa. The maximum reduction was at the full replacement of POC, which registered a value of 27% lower than the control concrete. During the visual examination of some broken specimens, as shown in Figure 16, the failure of the specimen was mainly due to the breaking of POC particles, while the bonding between the hardened cement paste and POC remained good. As such, the failure occurred through the POC coarse, which is weaker as compared to the concrete matrix and the aggregate-matrix interface. Meanwhile, at 28 days, the POCP concrete recorded an increase of between 10% and 31%, with respect to POC concrete mixes, while the splitting tensile values were in the range of 3.23 to 3.83 MPa. The development of the splitting tensile strength of the POCP concrete mixes up to 90 days is presented in Figure 17. All of the mixes had tensile strength values close to that of the control mix at different ages, and no trend was found linking this property with the replacement ratio of POC. The maximum strength reduction of the POCP concrete was at full replacement, which registered a value of 11% lower than the control mix. However, it became closer to the control at the age of 90 days as shown in Figure 18.

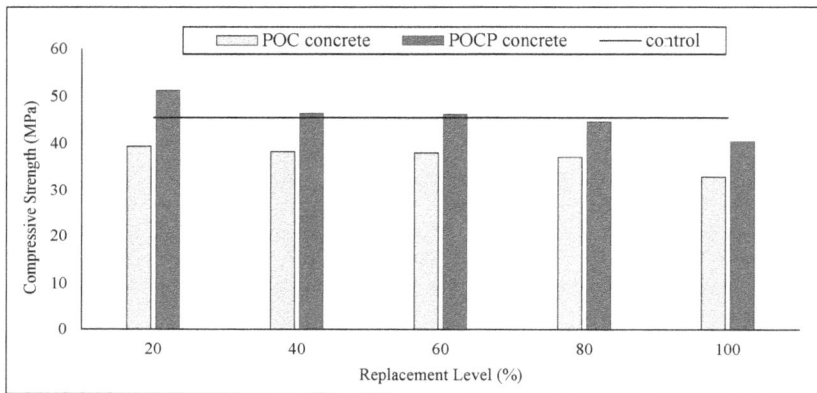

**Figure 12.** Development of POCP concrete strength.

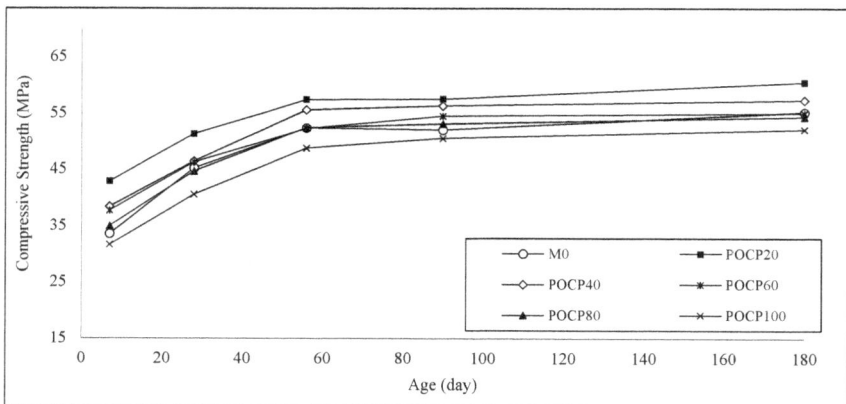

**Figure 13.** 28-day compressive strength of POC and POCP concretes.

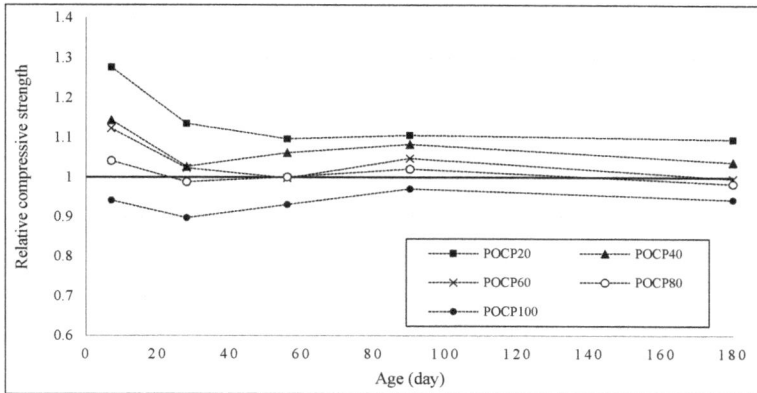

**Figure 14.** Relative compressive strength of POCP concrete.

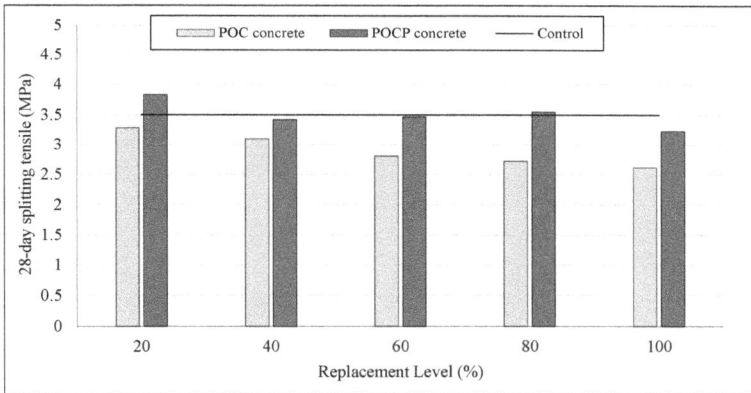

**Figure 15.** 28-day splitting tensile of POC and POCP concretes.

**Figure 16.** Splitting tensile failure.

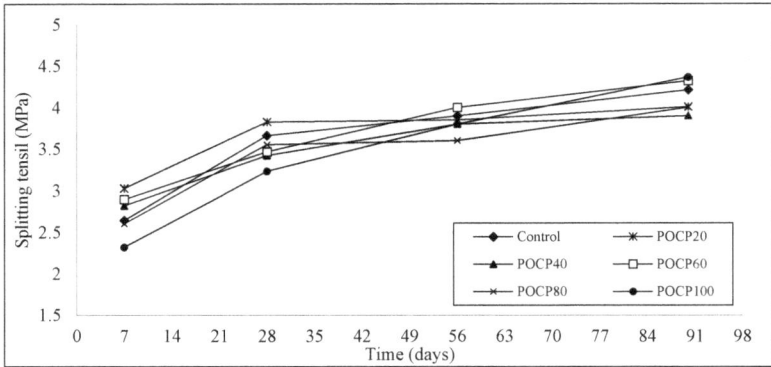

**Figure 17.** Development of splitting tensile of POCP concrete.

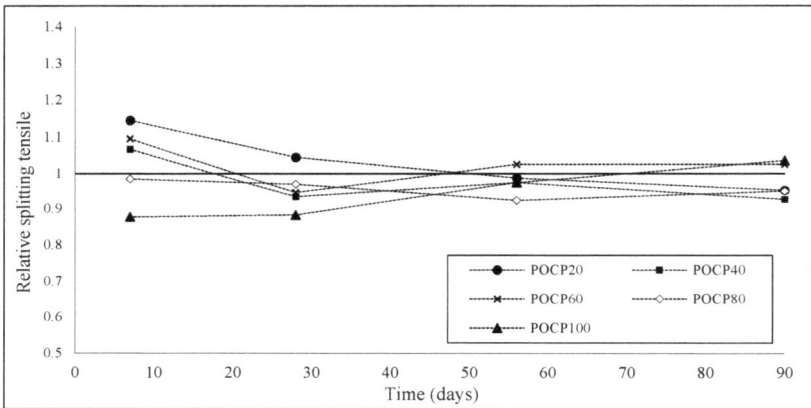

**Figure 18.** Relative splitting tensile strength of POCP concrete.

Nevertheless, the obtained tensile strength of the POCP concrete was always higher than the minimum recommended by ASTM C330 for structural lightweight concrete, which is 2 MPa. Previous studies [20–22] reported that the 28 days splitting tensile strength of lightweight concrete ranged between 1.10 and 2.41 MPa for moist cured concrete. In general, lightweight concrete with cube compressive strengths of 50, 40, and 30 MPa has a splitting tensile strength in the range of 2.5–3.8, 2.2–3.3, and 1.8–2.7 MPa, respectively [23]. However, the measured splitting tensile strength of POCP concrete in this study was in the range of 3.23–3.82 MPa at 28 days, which is higher than the values reported in previous studies on lightweight concrete.

The relationships between compressive strength, flexural, and splitting tensile strength at 28 days are tabulated in Table 5. In general, the splitting tensile strength for normal weight concrete ranges from 8% to 14% of its compressive strength [23]. The splitting/compressive strength ratio for normal weight concrete is higher when compared to the lightweight concrete [24]. Holm (2000) [25] reported that lightweight concrete that is moist cured has a splitting tensile strength of generally between 6% and 7% of its compressive strength. However, at the age of 28 days, the splitting tensile strength of the POCP concrete in this study ranged from 6.5% to 8% of the compressive strength. This is similar to the tensile/compressive strength ratio ranging from 6.6% to 9% of the lightweight concrete made with an artificial lightweight aggregate, as reported by Haque (2014) [24].

**Table 5.** Flexure, splitting and compressive strength relationship at 28 days.

| ID | Compressive Strength (MPa) $f_{cu}$ | Flexural Strength (MPa) $f_r$ | $\frac{f_r}{f_{cu}}$ | Splitting Tensile (MPa) $f_t$ | $\frac{f_t}{f_{cu}}$ |
|---|---|---|---|---|---|
| M0 | 45.16 | 4.422 | 9.79 | 3.663 | 8.11 |
| POCP20 | 51.27 | 4.521 | 8.82 | 3.829 | 7.47 |
| POCP40 | 46.39 | 4.501 | 9.70 | 3.422 | 7.16 |
| POCP60 | 46.22 | 4.381 | 9.48 | 3.315 | 6.82 |
| POCP80 | 44.63 | 4.800 | 10.76 | 3.553 | 7.96 |
| POCP100 | 40.52 | 4.534 | 11.19 | 3.233 | 7.51 |

A parabolic relationship with a correlation coefficient of 0.86 was observed between the 28-day compressive strength and splitting tensile of the POCP concrete, as shown in Figure 19. Figure 20 shows the comparison between the splitting tensile strength predicted by various equations listed in Table 6 and the experimental values. Equation (1) was proposed based on the results in this study for the POCP concrete. All equations with their descriptions are tabulated in Table 6. It can be seen that the equation of POCP concrete in this study is different from the equations proposed in previous studies due to the usage of a different type of aggregate in the mixtures and its physical properties. Additionally, Figure 20 revealed that the predicted splitting tensile from compressive strength of the POCP concrete is close and comparable with the Equations of (2) and (6), and overestimation than that of Equations (3)–(5).

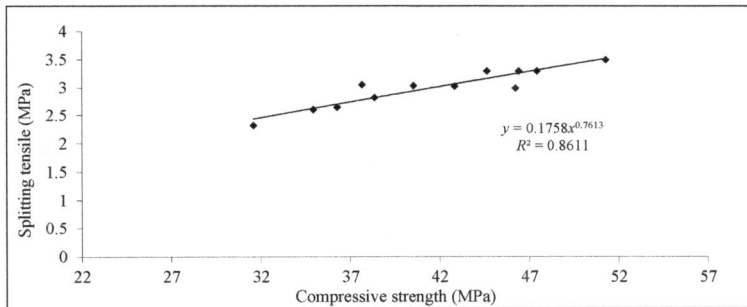

**Figure 19.** Splitting tensile and compressive strength relationship of POCP concrete.

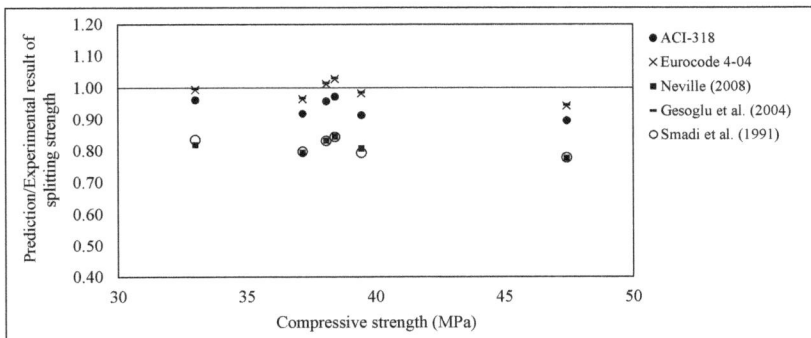

**Figure 20.** Experimental and theoretical 28-day splitting tensile of POCP concretes.

**Table 6.** Practical equations for splitting tensile strength of concrete.

| Equation | Description | Reference | Equation No. |
|---|---|---|---|
| $f_t = 0.176 f_{cu}^{0.76}$ | POCP concrete with cube compressive strength ranging between 40 and 51 MPa | This study | Equation (1) |
| $f_t = 0.53 f_{cy}^{0.5}$ | ACI-318-11 | [26] | Equation (2) |
| $f_t = 0.3 f_{cy}^{0.67}$ | Eurocode 4-04 | Eurocode 4-04 | Equation (3) |
| $f_t = 0.46 f_{cy}^{0.5}$ | From natural Tuff LWAC with a compressive strength as high as 60 MPa | [27] | Equation (4) |
| $f_t = 0.23 f_{cu}^{0.66}$ | For pelletized blast slag LWAC with cube compressive strength ranging from 10 to 65 MPa | [28] | Equation (5) |
| $f_t = 0.27 f_{cu}^{0.67}$ | Concrete with an artificial LWA has cube compressive strength ranging between 21 and 47 MPa | [29] | Equation (6) |

Note: $f_t$, splitting tensile; $f_{cu}$, cube compressive strength; $f_{cy}$, cylinder compressive strength.

## 6.5. Flexural Strength

A flexural strength test was carried out in accordance with BS1881: Part 118. The flexural strength results of the POC and POCP concrete, respectively, revealed that as the POC coarse content increased, the flexural strength decreased, as shown in Figure 21. All POC concrete have slightly lower flexural strength values when compared to that of the control concrete. At 28 days, the flexural strength of the POC concrete was in the range of 3.75 to 4.42 MPa. The maximum reduction was at full replacement, with approximately 15% lower than the control concrete. However, a significant increase in the flexural strength was achieved with the use of POCP in the POC concrete mixtures. The improvement was in the range of 5%–25% higher when compared to the POC concrete (pre-coating). Figure 22 illustrates the flexural strength development with curing age of POCP concrete mixes of up to 90 days. Flexural strength of the POCP concrete at different ages ranged from 4.01 to 6.15 MPa, and it was always higher than the control mix value at a specific age. Previous studies [20,21,30,31] also revealed that lightweight concrete has flexural strength in the range of between 2.13 and 4.93 MPa. Shetty (2005) [32] reported that for concrete with a compressive strength of 25 MPa and above, under continuous moist curing, the flexural strength is generally within the range of 8% to 11% of its compressive strength. Teo et al. (2006) [21] showed that the flexural strength varied in the range of 8% to 13% of the compressive strength. As stipulated in Table 5, the POCP concrete had a ratio in the range of 9.8–11.2 for flexural to compressive strength. This result is in good agreement with conventional values derived for concrete made with natural aggregates.

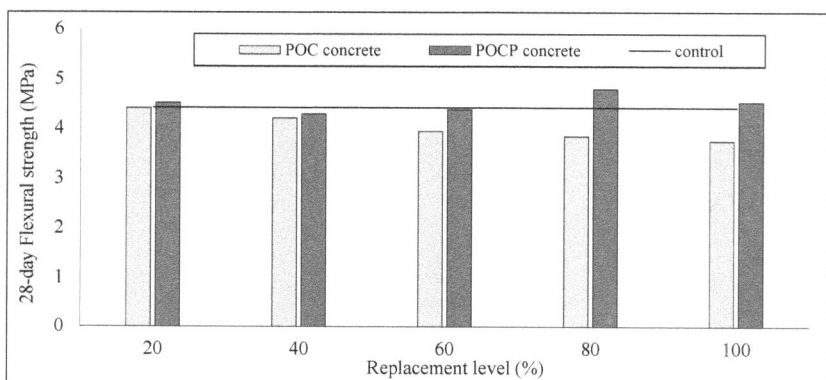

**Figure 21.** 28-day Flexural strength of POC and POCP concrete.

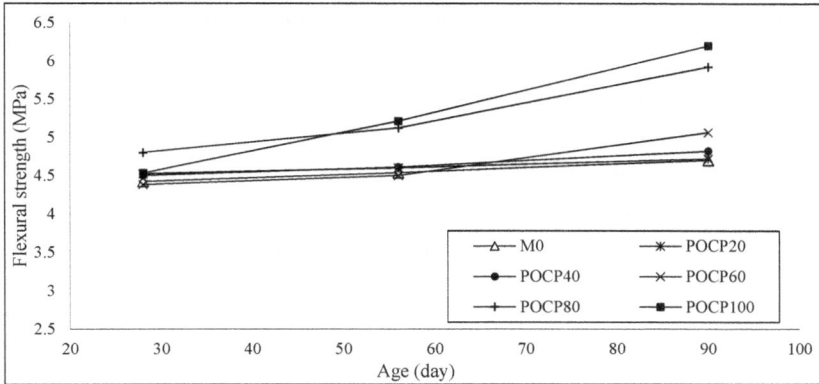

**Figure 22.** Developing of flexural strength of POCP concretes.

*6.6. Modulus of Elasticity*

Modulus of elasticity (MOE) was conducted according to BS 1881: Part 121 on the cylinder specimens with a dimension of 150 mm diameter × 300 mm length. The results of the MOE of the concrete specimen containing different replacement levels of POC coarse with and without POCP are shown in Figure 23. Both POC and POCP concrete had a 28-day MOE ranging between 22 and 32 GPa, and the 28-day compressive strength ranged between 33 and 51 MPa. The incorporation of POC coarse negatively affected the MOE values of concrete. The results revealed that the MOE of the POC concrete was 9% to 31% lower than the control mix. Often, the quality of the coarse aggregate greatly affects the elastic modulus. A comparison between the MOE values of the POC and POCP concrete before and after coating at 28 days showed that the addition of POCP had a significant effect on the MOE of the POC concrete. POCP concrete had a 28-day MOE range between 28 and 32 GPa, which is 14%–46% higher than that of POC concrete (pre-coating). Furthermore, the addition of POCP resulted in the decreasing of the water to powder ratio, which benefited the elastic modulus property. Krizova et al. (2004) [33] studied the influence of the water-binder ratio on the static modulus of concrete and reported that using a lower volume of mixing water reduced the number of cracks created by drying, and consequently improved the elastic modulus.

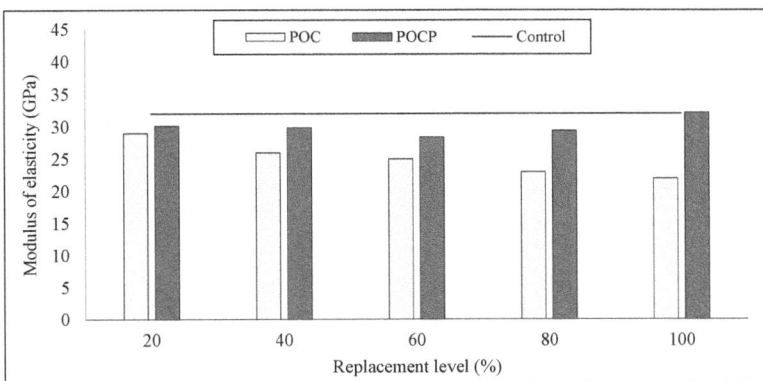

**Figure 23.** 28-day modulus of elasticity of POC and POCP concretes.

The increase in the MOE values of the POCP concrete with respect to the POC concrete mixes can also attributed to the enhancement of the interfacial transition zone. Domagała et al. (2011) [34] reported that for LWAC, a strong bond was formed between the cement matrix and aggregate due to the higher water absorption of LWA and its rough texture, resulting in a higher modulus of elasticity. Also, the rough surface texture further ensured that the bond between the surrounding hydrated cement paste and aggregate was better, thus improving the mechanical properties of the concrete [34]. The development of the MOE values of the POCP concrete up to 180 days is shown in Figure 24.

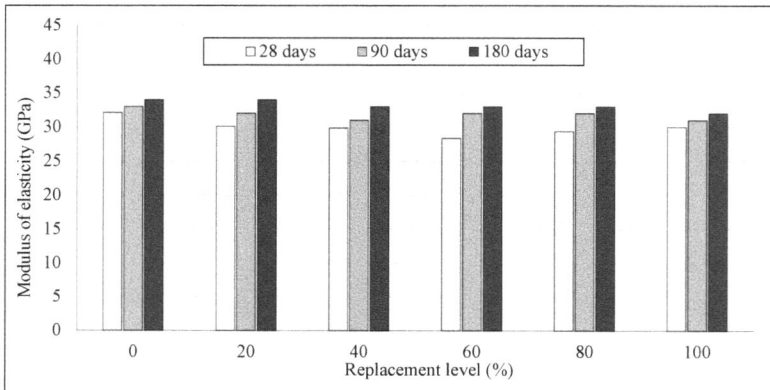

**Figure 24.** Development of modulus of elasticity of POCP concretes.

Various standards relate the MOE of concrete to its compressive strength and density. Figure 25 shows a comparison of the MOE value of the POCP concrete in this study with those predicted by various equations given in Table 7. The formulas presented in ACI 318 [35] defines this relationship in terms of either the square root of compressive strength or the combination of density and the square root of the compressive strength in Equations (8) and (9), respectively. This is applicable for density levels of 1440–2480 kg/m$^3$ and strength levels of 21–35 MPa. Hossain et al. (2011) [36] proposed Equation (11) based on data for lightweight concrete incorporating pumice with 28-day density ranging between 1460 and 2185 kg/m$^3$. Meanwhile, a cylinder compressive strength of 16–35 MPa. Equation (12) was proposed by Tasnimi (2004) [37], who presented information on artificial LWA concrete with a cylinder compressive strength of about 15–55 MPa.

**Table 7.** Practical Equations for Elastic Modulus of Concrete.

| Equation | Reference | Equation No. |
|---|---|---|
| $E_c = 0.0017 W_c^2 f_{cu}^{0.33}$ | [39] | Equation (7) |
| $E_c = 4730 f_{cy}^{0.5}$ | [35] | Equation (8) |
| $E_c = 0.043 W_c^{1.5} f_{cy}^{0.5}$ | [35] | Equation (9) |
| $E_c = 9500 f_{cy}^{0.33}$ | [40] | Equation (10) |
| $E_c = 0.03 W_c^{1.5} f_{cy}^{0.5}$ | [36] | Equation (11) |
| $E_c = 2.1684 f_{cy}^{0.535}$ | [37] | Equation (12) |
| $E_c = (0.062 + 0.0297 f_{cy}^{0.5}) W_c^{1.5}$ | [41] | Equation (13) |
| $E_c = 22,000 f_{cy}^{0.033}$ | [38] | Equation (14) |

Note: $E_c$, modulus of elasticity; $W_c$, concrete density; $f_{cu}$, cube compressive strength; $f_{cy}$, cylinder compressive strength.

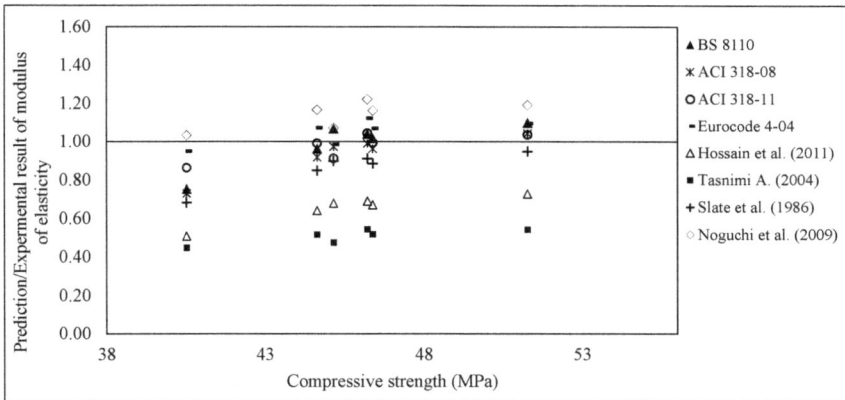

**Figure 25.** Experimental and theoretical of 28-day modulus of elasticity of POCP concretes.

As shown in Figure 25, among all of the equations, the MOE values of the POCP concrete at 28 days was close to and comparable with the values calculated using Equations (7), (8), and (10) as recommended by BS 8110, ACI 318 [35], and Noguchi et al. (2009) [38], respectively, for predicting the elastic modulus MOE of concrete in terms of its compressive strength and density. While the other equations underestimated the MOE values.

*6.7. Water Absorption*

The water absorption was carried out according to BSI 1881-122 [42]. Figure 26 shows the percentage of water absorption of the POC and POCP concrete specimens, as well as for the control concrete at 28 days. It is obvious that at the same w/c ratio of POC concrete mixes, the water absorption was higher than the control mix and tends to increase with the increasing volume of POC contents. At 28 days, the water absorption of the POC concrete was in the range of 35% to 80%, higher than that of the control mix. The water absorption has a direct relationship with the voids, the absorption increased as the voids increased [43]. A similar trend was observed in the results obtained by Teo et al. (2010) [44], which indicated that the high porosity of OPS aggregate increased the water absorption of the concrete as compared to the conventional concrete, like other lightweight aggregates concrete.

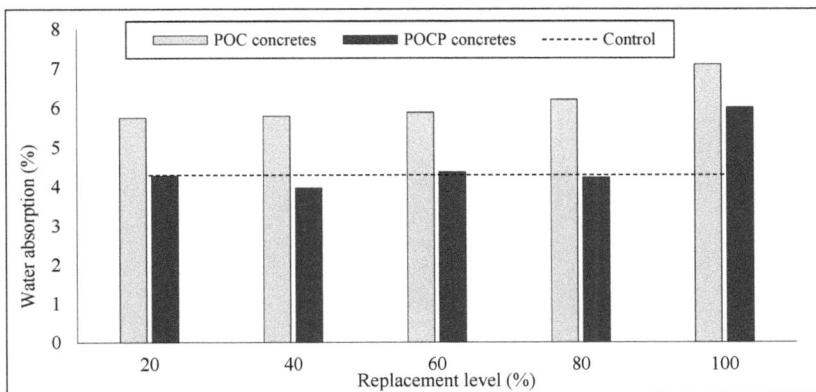

**Figure 26.** 28-day water absorption of POC and POCP concretes.

The lower porosity of the granite aggregate in the normal concrete mixture restricts the rate of water absorption as compared to the POC concrete. Most artificial lightweight concrete exhibits significantly higher water absorption than normal weight concrete [20]. Topcu (1997) [45] is of the idea according to the result of his own study that there is a parabolic connection between water absorption and concrete density, "the lower the concrete density, the higher water absorption capacity". However, the values of water absorption of the POCP concrete mixes are comparable to the natural aggregate concrete. The addition of POCP resulted in a decrease in the value of water absorption. At 28 days, the reduction was in the range of 15% to 32% with respect to the POC concrete. The increase of POCP content was advantageous to the concrete and resulted in a more condensed microstructure. The low water absorption of the POCP mixes is also attributed to the denser interfacial zone between the aggregate and mortar matrix with respect to that of the POC concrete mixes.

Moosberg et al. (2004) [46] reported that the physical effect of the mineral admixture on the concrete properties occurred as a result of pervading the fillers into the void between the cement particles. Therefore, the incorporation of POCP has a similar effect to the mineral admixtures in the concrete, by reducing the pore size, which resulted in highly densified paste. As such, the concrete water absorption decreased. It is also obvious that SP played an important role in enhancing the fluidity of the POCP concrete mixes and maximized the compaction subsequently resulting in the production of a high impermeable concrete. Figure 27 shows the percentage of water absorption of the POCP concrete specimens as well as for the control concrete subjected to 7, 28, 90, and 180 days of moist curing after demolding.

*6.8. Drying Shrinkage*

A demountable mechanical strain gauge (DEMEC), with a precision of 1 μm, was used to monitor the total liner shrinkage. The DEMEC was placed over two steel studs at a 200 mm gauge length, which had been glued onto the three as cast surfaces. The development of the shrinkage strains with a drying period of up to 180 days under an initial water curing condition of seven days is shown in Figure 28. The specimens were exposed to uncontrolled laboratory conditions, with humidity ranging between 60% and 85%, and temperature ranging between 26 and 35 °C. The test results showed that the POCP concrete mixes has a lower drying shrinkage strain when compared to the control concrete. Furthermore, the addition of POCP significantly improved the drying shrinkage of the POC concrete. The increase in POCP content resulted in decreased drying shrinkage. Also, it was observed that the mixes with a higher content of POCP have lower a drying shrinkage.

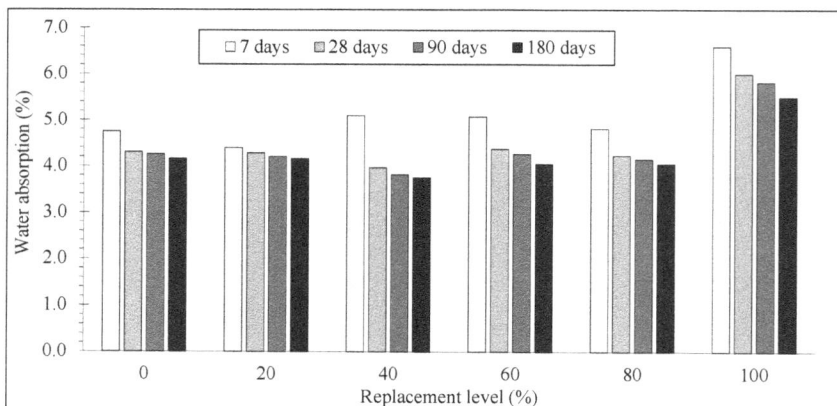

**Figure 27.** Water absorption of POCP concrete at different ages.

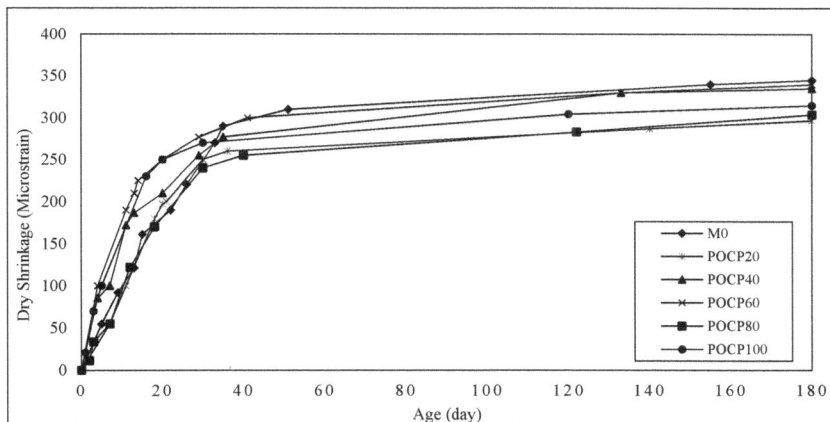

**Figure 28.** Dry shrinkage for POCP concrete.

The lower drying shrinkage values of the POCP concrete with respect to the control concrete can be attributed to two main factors. Firstly, the incorporation of POCP reduced the pore sizes in concrete. The transformation of large pores to fine pores decreased the evaporation of water from the concrete surface, and hence, reduced the drying shrinkage strain [47]. Secondly, the quality of cement paste directly influenced the drying shrinkage of the concrete such that the dry shrinkage increased with the increase in water content of the paste [48]. POCP concrete mixes have lower water to powder ratio when compared to the control concrete, which would be expected to cause a lower drying shrinkage strain. In general, it was observed that the difference between all of the mixes shrinkage values was not significant at a specific age.

*6.9. Chloride Permeability*

According to the ASTM C1202 test, a concrete specimen of 50 mm thick and 100 mm diameter is subjected to a 60 V applied DC voltage for 6 h using the apparatus shown in Figure 29. 3.0% NaCl solution was filled in one reservoir while in the other reservoir is a 0.3 M NaOH solution. The total charge passed was determined and this was used to rate the concrete according to the criteria given in Table 8.

Rapid Chloride Permeability Test (RCPT) was conducted to investigate the performance of the concrete mixes against chloride ingress. The lower the total charge passed through the concrete matrix, the higher the resistance to chloride penetration. RCPT was conducted for the mixes containing a replacement of 20%, 60%, and 100% of POC coarse before and after coating as well as for the control concrete. The charged passed was obtained by measuring an average of three samples at the ages of 28, 90, and 180 days. According to the ASTM rating standard, the control concrete suffered high chloride-ion penetrability at the ages of 28 and 90 days since the charge passed was higher than 4000 coulombs. Meanwhile, at the age of 180 days, the control concrete showed moderate chloride-ion penetrability. At 28 days, a slight variation in the total charge passed with respect to the percentage of POC coarse replacement was observed, as shown in Figure 30. The chloride ion resistance of the POC concrete mixes was similar and comparable to the control concrete. The total charge that passed through the POC concrete ranged between 4331 and 4895 columns, falling in the range of high chloride penetrability. Similar to the study by Chia (2002) [49] on LWA concrete, the results of the RCPT indicated that the electric charge passed through the LWC was in the same order as those through the corresponding normal weight concrete (NWC). Furthermore, the water to cement ratio is one of the main factors affecting the total charge passing through the concrete specimens. All of the POC concrete, including the control mix, had the same water to cement ratio of 0.53. Therefore, the total charge passed values are expected to be similar for all of the specimens at a specific age. Shi (2003) [50] reported that the water to powder ratio ranging between 0.4 and 0.5 for conventional concrete can achieve a charge passed of 2000 to 4000 coulombs, which is indicated as Moderate. Meanwhile, from the results shown in Figure 31, a progressive reduction was observed in the chloride penetrability from the POC to POCP concrete, respectively. Specimens containing additional POCP exhibited a greater chloride-ion resistance as compared to POC mixes as well as to the control concrete. The chloride ion penetration exhibited a higher reduction for the mixes with a higher content of POCP. At 28 days, the charge passed of the POCP concrete was in the range of 12% to 70% lower than that of the POC concrete. The reason for the significant reduction in chloride ion penetration was due to the low water to powder ratio of the mixes, which made the concrete more densify. Chia (2002) [49] reported that the resistance of concrete to chloride penetration increased with the reduction of w/cm (water to cementitious materials ratio). At 90 and 180 days, the effect of this was even more favorable. POCP concrete exhibited a general downward trend in the amount of electrical charge passed as the age increased.

**Figure 29.** Concrete specimens and test set up (ASTM C1202).

**Table 8.** RCPT ratings (per ASTM C1202).

| Charge Passed (Coulombs) | Chloride Ion Penetrability |
|---|---|
| >4000 High | High (H) |
| 2000–4000 Moderate | Moderate (M) |
| 1000–2000 Low | Low (L) |
| 100–1000 Very Low | Very Low (V.L) |

It is well known that the use of supplementary cementing materials improves pore structure and reduces the permeability of hardened concrete [51]. Smith (2006) [52] presented data that showed that the conductivity of the pore solution could be lowered with the use of mineral admixtures such as ground granulated blast furnace slag (GGBS), fly ash, and silica fume. One of the most important factors affecting the permeability of concrete is the internal pore structure, which in turn is dependent on the extent of the hydration of the cementitious materials [53]. Uysal et al. (2012) [54] reported that the factor that affects the pore system of concrete is the filler effect, which influences the total volume and size distribution of pores, and finally affecting the concrete permeability. Therefore, the incorporation of POCP had similar effect to the mineral admixtures in the concrete by reducing the pore size, which resulted in a highly densified paste and thus lowered the conductivity of the pore solution. The addition of POCP resulted in an increased powder material which is advantageous to the permeability resistance of the ions. Furthermore, the low permeability of chlorides for the POCP concrete can be justified by the low water to powder ratio in the POCP concrete mixes when compared to POC mixes, as well as to the control concrete.

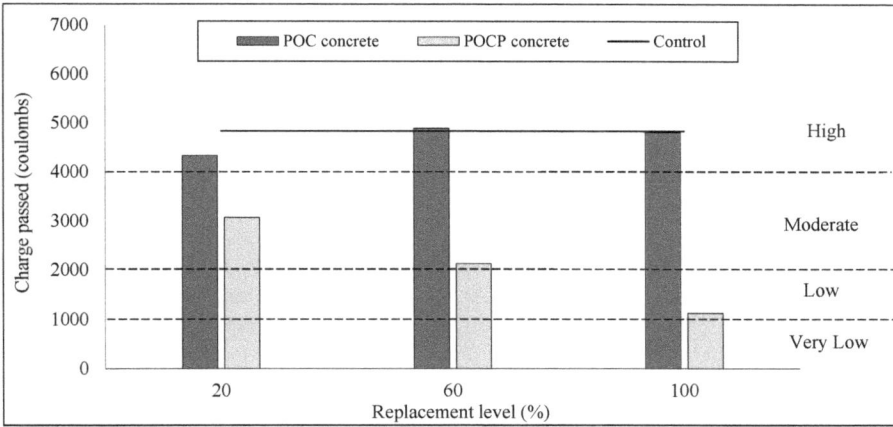

**Figure 30.** 28-day charge passed coulombs value of POC and POCP concretes.

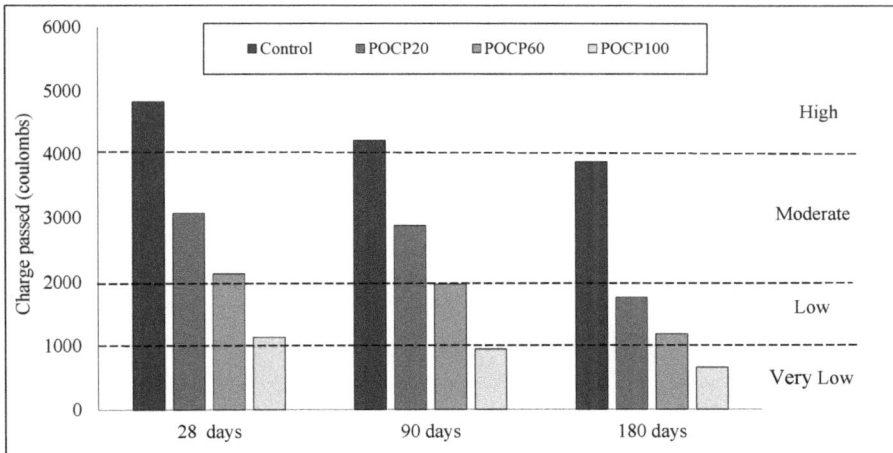

**Figure 31.** Charge passed coulombs value of POCP concretes.

## 7. Conclusions

Based on the experimental results of this work, the following conclusions can be drawn:

- POC, being highly porous had a negative effect on the fresh and hardened concrete properties when the coarse aggregate is substituted with POC. However, incorporating additional POCP into the POC concrete mixes resulted in increasing the paste content required to make the mixes more cohesive. POCP together with SP proved to be beneficial to the workability of the POC concrete mixes.
- In general, there was reduction in the compressive strength when POC coarse aggregate was used. The concrete strength was lower when a higher content of conventional aggregate was substituted with the POC. At 28 days, the compressive strength of the POC concrete obtained was in the range between 33.01 and 39.32 MPa. Meanwhile, a significant reduction in compressive strength was avoided when POC coarse was coated using POCP as a filler material to the surface

voids. Thus, the compressive strength of the POCP concrete increased by 20% to 30% compared to the mixes before coating.

- Splitting tensile strength results of the POC concrete generally showed a trend similar to that observed in the compressive strength. The higher the contents of POC coarse, the lower the splitting tensile value. At 28 days, the splitting tensile of the POC concrete was in the range of 2.61 to 3.28 MPa. The maximum reduction was at full replacement of POC, which registered a value of 27% lower than the control concrete. Meanwhile, the POCP concrete recorded an increased ranging between 10% and 31% with respect to POC concrete mixes (pre-coating).

- All of the POC concrete mixes had slightly lower flexural strength values when compared to that of the control concrete. At 28 days, the flexural strength of the POC concrete was in the range of 3.75 to 4.42 MPa. The maximum reduction was at full replacement with approximately 15% lower than the control concrete. However, a significant increase in flexural strength was achieved when POCP was used in the POC concrete mixtures. The improvement was in the range of 5%–25% higher when compared to the POC concrete.

- Incorporation of the POC coarse negatively affected the MOE value of the concrete. The MOE of the POC concrete dropped by 9% to 31% lower than that of the control concrete. However, the POCP concrete had a 28-day MOE values range of between 28 and 32 GPa which was 14%–46% higher than that of the POC concrete (pre-coating).

- POC concrete mixes had higher water absorption when compared to the control mix and tends to rise with an increasing POC coarse contents. However, the addition of POCP resulted in a decrease in the value of water absorption when compared to the POC concrete by reducing the pore size, which resulted in highly densified paste.

- Specimens containing additional POCP exhibited a greater chloride-ion resistance as compared to POC mixes, as well as to the control concrete. At 28 days, the charge passed of the POCP concrete was in the range of 12% to 70%, lower than that the POC concrete.

- The results revealed that coating the surface voids of POC coarse with POCP significantly improved the engineering properties as well as the durability performance of the POC concrete. Thus, using POC as an aggregate and filler material may reduce the continuous exploitation of aggregates from primary sources. Also, this approach offers an environmentally friendly solution to the ongoing problems of palm oil waste material.

**Acknowledgments:** The authors would like to express their sincere thanks to the Ministry of Education (MOE), Malaysia for the support given through the research grant UM.C/625/1/HIR/MOHE/ENG/56 and Postgraduate Research Grant (PPP)—PG277-2015B.

**Author Contributions:** Fuad Abutaha designed and performed the experiments under the guidance and supervision of Hashim Abdul Razak. Hussein Adebayo Ibrahim assisted in the experimental works. The manuscript was written by Fuad Abutaha and revised by Hashim Abdul Razak.

**Conflicts of Interest:** The authors declare no conflict of interest.

## Appendix A

Detailed calculation of determine the required POCP using PP method:

1. The void volume ($V_{Void}$) using PP method for all the POC concrete mixes is determined by adopting Equations (A1) and (A2). This is represented by the volume of water ($V_{PPwater}$) required to fill up the PP container for different substitution levels of POC coarse as well as for the control mix. Please change the following equations to be Editable state.

$$V_{Void}(\mathrm{m}^3) = V_{PPwater} \tag{A1}$$

$$V_{Void}(\mathrm{m}^3/\mathrm{m}^3) = \frac{V_{PPwater}}{V_{container}} = \frac{Column\,A}{Column\,B} \tag{A2}$$

2. Equation (A3) is adopted to determine the percentage of void increment resulting from each replacement level of POC coarse in which the PP control mix void volume ($V_{PPcontrolVoid}$) is taken as a benchmark.

$$Void\ Increment\ (\%) = \frac{V_{void} - V_{PPcontrolVoid}}{V_{PPcontrolVoid}} \times 100 \tag{A3}$$

3. Adjust the control paste volume using Equation (A4), multiplying the PP paste volume by the correction factor:

$$V_{ControlPaste} = CorrectionFactor \times V_{PPcontrolPaste} \tag{A4}$$

4. The total paste volume required for different substitution level of POC concrete ($V_{TotalPaste}$) is calculated using (Equation A5) by increasing the control cement paste volume ($V_{ControlPaste}$) with the percentage of void increment obtained in Equation (A3).

$$V_{TotalPaste} = V_{ControlPaste} + (\% of\ Void\ Increment) \times V_{ControlPaste} \tag{A5}$$

5. The volume of the additional POCP ($V_{POCP}$) is obtained using Equation (A6) by deducting the control paste volume ($V_{ControlPaste}$) from the total paste volume ($V_{TotalPaste}$) obtained in Equation (A5). Equation (A7) represents the additional POCP in kg/m$^3$.

$$V_{POCP} = V_{TotalPaste} - V_{ControlPaste} \tag{A6}$$

$$POCP\ (\text{kg/m}^3) = V_{POCP} \times SG_{POCP} \tag{A7}$$

where $SG_{POCP}$ is Specific Gravity of POCP.

**Table A1.** POCP required for each substitution levels of POC coarse.

| Column | A | B | C | D | E | F | G | K | L |
|---|---|---|---|---|---|---|---|---|---|
| ID | PP Water Vol. (m³) | PP Container Vol. (m³) | Void Vol. (m³/m³) $\frac{A}{B}$ | Void Increment (%) $\frac{C-0.3049}{0.3049} \times 100$ | Control Mix Correction Factor | Control Paste Vol. (m³/m³) $0.3049 \times 1.16$ | Total Paste Vol. (m³/m³) $(D \times 0.356) + 0.356$ | POCP Vol. Required (m³/m³) G-0356 | POCP (kg/m³) $K \times POCP_{SG}$ |
| M0 | 0.00057 | 0.0018696 | 0.3049 | – | 1.16 | 0.356 | 0.356 | – | – |
| POCP20 | 0.00062 | 0.0018696 | 0.3289 | 7.92 | – | – | 0.384 | 0.028 | 70 |
| POCP40 | 0.00063 | 0.0018696 | 0.337 | 10.55 | – | – | 0.393 | 0.037 | 93 |
| POCP60 | 0.00064 | 0.0018696 | 0.3423 | 12.31 | – | – | 0.399 | 0.043 | 108 |
| POCP80 | 0.00067 | 0.0018696 | 0.3584 | 17.57 | – | – | 0.418 | 0.062 | 156 |
| POCP100 | 0.0007 | 0.0018696 | 0.3744 | 22.84 | – | – | 0.437 | 0.081 | 203 |

## References

1. Rashid, M.A.; Salam, M.A.; Shill, S.K.; Hasan, M.K. Effect of replacing natural coarse aggregate by brick aggregate on the properties of concrete. *Dhaka Univ. Eng. Technol. J.* **2012**, *1*, 17–22.
2. Katz, A. Properties of concrete made with recycled aggregate from partially hydrated old concrete. *Cem. Concr. Res.* **2003**, *33*, 703–711. [CrossRef]
3. Alnahhal, M.F.; Alengaram, U.J.; Jumaat, M.Z.; Alqedra, M.A.; Mo, K.H.; Sumesh, M. Evaluation of industrial by–products as sustainable pozzolanic materials in recycled aggregate concrete. *Sustainability* **2017**, *9*, 767. [CrossRef]
4. Mannan, M.; Neglo, K. Mix design for oil–palm–boiler clinker (OPBC) concrete. *J. Sci. Technol.* **2010**, *30*, 111–118. [CrossRef]
5. Hosseini, S.E.; Wahid, M.A. Utilization of palm solid residue as a source of renewable and sustainable energy in malaysia. *Renew. Sustain. Energy Rev.* **2014**, *40*, 621–632. [CrossRef]
6. Halimah, M.; Tan, Y.A.; Nik Sasha, K.K.; Zuriati, Z.; Rawaida, A.I.; Choo, Y.M. Determination of life cycle inventory and greenhouse gas emissions for a selected oil palm nursery in malaysia: A case study. *J. Oil Palm Res.* **2013**, *25*, 343–347.

7.  Kanadasan, J.; Razak, H.A. Mix design for self–compacting palm oil clinker concrete based on particle packing. *Mater. Des.* **2014**, *56*, 9–19. [CrossRef]
8.  Fuad Abutaha, H.A.R. Jegathish Kanadasan. Effect of palm oil clinker (POC) aggregates on fresh and hardened properties of concrete. *Constr. Build. Mater.* **2016**, *112*, 416–423. [CrossRef]
9.  Kanadasan, J.; Fauzi, A.F.A.; Razak, H.A.; Selliah, P.; Subramaniam, V.; Yusoff, S. Feasibility studies of palm oil mill waste aggregates for the construction industry. *Materials* **2015**, *8*, 6508–6530. [CrossRef] [PubMed]
10. Abdullahi, M.; Al-Mattarneh, H.; Hassan, A.A.; Hassan, M.; Mohammed, B. Trial mix design methodology for palm oil clinker (POC) concrete. In Proceedings of the International Conference on Construction and Building Technology, Kuala Lumpur, Malaysia, 16–20 June 2008.
11. Ibrahim, H.A.; Razak, H.A. Effect of palm oil clinker incorporation on properties of pervious concrete. *Constr. Build. Mater.* **2016**, *115*, 70–77. [CrossRef]
12. Kanadasan, J.; Abdul Razak, H. Utilization of palm oil clinker as cement replacement material. *Materials* **2015**, *8*, 8817–8838. [CrossRef] [PubMed]
13. Karim, M.R.; Hashim, H.; Razak, H.A. Assessment of pozzolanic activity of palm oil clinker powder. *Constr. Build. Mater.* **2016**, *127*, 335–343. [CrossRef]
14. Mangulkar, M.; Jamkar, S. Review of particle packing theories used for concrete mix proportioning. *Int. J. Sci. Eng. Res.* **2013**, *4*, 143–148.
15. Koehler, E.P. *Aggregates in Self-Consolidating Concrete*; Final Report; International Center for Aggregate Research (ICAR) Report; University of Texas: Austin, TX, USA, 2007.
16. Kanadasan, J.; Razak, H.A. Engineering and sustainability performance of self–compacting palm oil mill incinerated waste concrete. *J. Clean. Prod.* **2015**, *89*, 78–86. [CrossRef]
17. Abutaha, F.; Abdul Razak, H.; Kanadasan, J. Effect of palm oil clinker (POC) aggregates on fresh and hardened properties of concrete. *Constr. Build. Mater.* **2016**, *112*, 416–423. [CrossRef]
18. Ibrahim, H.A.; Abdul Razak, H.; Abutaha, F. Strength and abrasion resistance of palm oil clinker pervious concrete under different curing method. *Constr. Build. Mater.* **2017**, *147*, 576–587. [CrossRef]
19. Tuan, B.L.A.; Hwang, C.-L.; Lin, K.-L.; Chen, Y.-Y.; Young, M.-P. Development of lightweight aggregate from sewage sludge and waste glass powder for concrete. *Constr. Build. Mater.* **2013**, *47*, 334–339. [CrossRef]
20. Mannan, M.A.; Ganapathy, C. Engineering properties of concrete with oil palm shell as coarse aggregate. *Constr. Build. Mater.* **2002**, *16*, 29–34. [CrossRef]
21. Teo, D.; Mannan, M.; Kurian, V. Structural concrete using oil palm shell (OPS) as lightweight aggregate. *Turk. J. Eng. Environ. Sci.* **2006**, *30*, 251–257.
22. Abdullah, A. Palm oil shell aggregate for lightweight concrete. In *Waste Materials Used in Concrete Manufacturing*; Chandra, S., Ed.; Noyes Publ.: Westwood, NJ, USA, 1996; pp. 624–636.
23. Shafigh, P.; Jumaat, M.Z.; Mahmud, H.B.; Hamid, N.A.A. Lightweight concrete made from crushed oil palm shell: Tensile strength and effect of initial curing on compressive strength. *Constr. Build. Mater.* **2012**, *27*, 252–258. [CrossRef]
24. Haque, M.; Al-Khaiat, H.; Kayali, O. Strength and durability of lightweight concrete. *Cem. Concr. Compos.* **2004**, *26*, 307–314. [CrossRef]
25. Holm, T.A.; Bremner, T.W. *State of the Art Report on High-Strength, High-Durability Structural Low-Density Concrete for Applications in Severe Marine Environments*; US Army Corps of Engineers, Engineer Research and Development Center: Vicksburg, MS, USA, 2000.
26. ASTM International. *Standard Test Method for Splitting Tensile Strength of Cylindrical Concrete Specimens*; ASTM International: West Conshohocken, PA, USA, 2011.
27. Smadi, M.; Migdady, E. Properties of high strength tuff lightweight aggregate concrete. *Cem. Concr. Compos.* **1991**, *13*, 129–135. [CrossRef]
28. Neville, A.M. *Properties of Concrete*, 14th ed.; Prentice Hall: Upper Saddle River, NJ, USA, 2008.
29. Gesoğlu, M.; Özturan, T.; Güneyisi, E. Shrinkage cracking of lightweight concrete made with cold-bonded fly ash aggregates. *Cem. Concr. Res.* **2004**, *34*, 1121–1130. [CrossRef]
30. Okafor, F.O. Palm kernel shell as a lightweight aggregate for concrete. *Cem. Concr. Res.* **1988**, *18*, 901–910. [CrossRef]
31. Mahmud, H. Ductility behaviour of reinforced palm kernel shell concrete beams. *Eur. J. Sci. Res.* **2008**, *23*, 406–420.
32. Shetty, M.S. *Concrete Technology Theory and Practice*; S. Chand & Company Ltd.: New Delhi, India, 2005.

33. Krizova, K.; Hela, R. Selected Technological Factors Influencing the Modulus of Elasticity of Concrete. *World Acad. Sci. Eng. Technol. Int. J. Civ. Environ. Struct. Constr. Archit. Eng.* **2014**, *8*, 593–595.

34. Domagała, L. Modification of properties of structural lightweight concrete with steel fibres. *J. Civ. Eng. Manag.* **2011**, *17*, 36–44. [CrossRef]

35. *ACI 318-08 Building Code Requirements for Structural Concrete and Commentary;* ACI Standard; American Concrete Institute: Farmington Hills, MI, USA, 2008; p. 465.

36. Hossain, K.; Ahmed, S.; Lachemi, M. Lightweight concrete incorporating pumice based blended cement and aggregate: Mechanical and durability characteristics. *Constr. Build. Mater.* **2011**, *25*, 1186–1195. [CrossRef]

37. Tasnimi, A. Mathematical model for complete stress-strain curve prediction of normal, light-weight and high-strength concretes. *Mag. Concr. Res.* **2004**, *56*, 23–34. [CrossRef]

38. Noguchi, T.; Tomosawa, F.; Nemati, K.M.; Chiaia, B.M.; Fantilli, A.P. A practical equation for elastic modulus of concrete. *ACI Struct. J.* **2009**, *106*, 690.

39. *BS 8110: Part 2, Structural Use of Concrete. Part 2: Code of Practice for Special Circumstances;* British Standards Institution: London, UK, 1985.

40. Australian Standard. *General Purpose and Blended Cements;* Australian Standard: Sydney, Australia, 2014.

41. Nilson, A.H.; Martinez., S. Mechanical properties of high-strength lightweight concrete. *Aci Mater. J.* **1991**, *88*, 240–247.

42. *BSI BS 1881: Part 122. Method for Determination of Water Absorption;* British Standard Institution: London, UK, 1983.

43. Wongkeo, W.; Thongsanitgarn, P.; Ngamjarurojana, A.; Chaipanich, A. Compressive strength and chloride resistance of self-compacting concrete containing high level fly ash and silica fume. *Mater. Des.* **2014**, *64*, 261–269. [CrossRef]

44. Teo, D.; Mannan, M.; Kurian, V. Durability of lightweight OPS concrete under different curing conditions. *Mater. Struct.* **2010**, *43*, 1–13. [CrossRef]

45. Topcu, I.B. Physical and mechanical properties of concretes produced with waste concrete. *Cem. Concr. Res.* **1997**, *27*, 1817–1823. [CrossRef]

46. Moosberg-Bustnes, H.; Lagerblad, B.; Forssberg, E. The function of fillers in concrete. *Mater. Struct.* **2004**, *37*, 74–81. [CrossRef]

47. Tangchirapat, W.; Jaturapitakkul, C.; Chindaprasirt, P. Use of palm oil fuel ash as a supplementary cementitious material for producing high-strength concrete. *Constr. Build. Mater.* **2009**, *23*, 2641–2646. [CrossRef]

48. Ahmad, M.H.; Mohd Noor, N.; Adnan, S.H. Shrinkage of Malaysian palm oil clinker concrete. In Proceedings of the International Conference on Civil Engineering Practice (ICCE08), Kuantan, Pahang, 12–14 May 2008.

49. Chia, K.S.; Zhang, M.-H. Water permeability and chloride penetrability of high-strength lightweight aggregate concrete. *Cem. Concr. Res.* **2002**, *32*, 639–645. [CrossRef]

50. Shi, C. *Another Look at the Rapid Chloride Permeability Test (Astm C1202 or Asshto T277);* FHWA Resource Center: Baltimore, MD, USA, 2003.

51. Stanish, K.D.; Hooton, R.D.; Thomas, M.D.A. *Testing the Chloride Penetration Resistance of Concrete: A Literature Review;* University of Toronto: Toronto, ON, Canada, 2000.

52. Smith, D. The Development of a Rapid Test for Determining the Transport Properties of Concrete. Master's Thesis, University of New Brunswick, Fredericton, NB, Canada, November 2006.

53. Joshi, P.; Chan, C. Rapid chloride permeability testing. *Concr. Constr.* **2002**, *47*, 37–43.

54. Uysal, M.; Yilmaz, K.; Ipek, M. The effect of mineral admixtures on mechanical properties, chloride ion permeability and impermeability of self-compacting concrete. *Constr. Build. Mater.* **2012**, *27*, 263–270. [CrossRef]

*coatings*

MDPI

*Article*

# Impact Wear of Structural Steel with Yield Strength of 235 MPa in Various Liquids

**Yueting Liu [1,2,*] and G.C.A.M. Janssen [2]**

[1] Materials Innovation Institute M2i, Elektronicaweg 25, 2628 XG Delft, The Netherlands
[2] Department of Precision and Microsystems Engineering, Delft University of Technology, Mekelweg 2, 2628 CD Delft, The Netherlands; G.C.A.M.Janssen@tudelft.nl
* Correspondence: y.liu-4@tudelft.nl; Tel.: +31-152-781-940

Academic Editor: Alicia Esther Ares
Received: 22 August 2017; Accepted: 20 November 2017; Published: 20 December 2017

**Abstract:** The wear of pipelines, used in slurry transport, results in high costs for maintenance and replacement. The wear mechanism involves abrasion, corrosion, impact, and the interaction among them. In this work, we study the effect of impact on the wear mechanism and wear rate. Results show that when the effect of impact is small, the wear mechanism is dominated by electrochemically induced surface modification, which leads to a lower wear rate in a corrosive environment than in a non-corrosive environment. By contrast, when the effect of impact is large, the wear mechanism is drastically altered. In that regime plastic deformation is important. The influence of corrosion in the high impact regime can be neglected. Our findings show the importance of including impact effect in the distinction of wear of slurry pipes.

**Keywords:** wear; corrosion; impact; deformation; surface modification

## 1. Introduction

Dredging is involved in keeping waterways navigable or for the purposes of constructing new land in freshwater or seawater areas. In dredging engineering, the sedimented sands or other solids, mixed with water, need to be transported by means of pipelines. The pipelines wear due to the interaction among erosion, abrasion, and corrosion, resulting in high costs for maintenance and replacement of the pipelines [1–3].

Researchers have studied and reported the wear of pipelines for decades [4–10]. Truscott [11] reviewed the research findings of 20 years before 1972 and summarized three determinant factors of wear: the properties of the slurry, the regime of the flow and the materials. The properties of slurry mainly include the particle hardness, size, shape (sharpness), specific mass, concentration. The mode of flow in principle determines the particle dispersion and particle motion, which eventually determine the wear mechanism. The properties of materials mainly include chemical composition, microstructure and hardness. The erosion wear of slurry pipelines results from two mechanisms, particle impact and scouring, where the latter occurs as a result of a sliding abrasive wear. Several apparatuses have been introduced to study the wear mechanism of slurry pipes [12,13]. These apparatuses aim to study the influence of multiple factors like the flow concentration, the particle size, or the flow velocity.

However, the material surface change due to corrosion is rarely reported, yet extremely important. In a corrosive environment, especially electrochemical corrosion, the surface of the material changes and the change could potentially alter the original material properties like hardness, and therefore it influences the final wear rate. In a previous paper [14], the authors reported the micro-coupling effect, occurring in a multiphase material in an electrochemically corrosive environment. In that study, a pearlitic steel, consisting of ferrite and cementite, was exposed in seawater while subject to abrasive wear. The coupling between ferrite and cementite, due to their electrochemical potential difference, leads to the dissolution of ferrite and the protruding of cementite, and subsequently the redistribution of cementite due to abrasion. The altered surface is much harder than the original due to the enrichment of harder phase: cementite.

In actual slurry transport, however, no noticeable seawater wear rate decrease was observed and reported. In this study, impact is introduced to identify the wear mechanism. Various liquids are used to provide different corrosive environments. We will argue that impact dominates in slurry transport. This domination of impact wear explains the negligible difference in wear in seawater and fresh water.

## 2. Experimental

### 2.1. Material Preparation

The material, used in this study, is a structural steel: S235 [15]. This material, with minimum yield strength of 235 MPa and hardness of 183 $HV_{0.1}$ ($\pm$9.2), is widely used in dredging industry for its good combination of mechanical properties and welding properties. The sample used in this study are cylinders, 30 mm in diameter and around 8 mm in thickness. The preparation procedure consists of three stages. First, the sample was grinded with silicon carbide sandpaper from 80 to 2400 mesh (particle size equals roughly 10 μm). Then the grinded sample was polished with diamond containing polishing liquid from 3 μm to 1 μm until the surface was mirror-like. Finally, the polished sample was cleaned in acetone with ultrasonic, followed by rinsing with distilled water and dried with room temperature air.

### 2.2. Experimental Procedure with a Hammering Pin on Disc

A modified pin on disc, as shown in Figure 1, was used to perform experiments. A hammering module was incorporated. The module contains a retractable component, powered by compressed air to hit and lift the pin. The impact height can be measured during experiments by the pin on disc software (TriboX 1.0, CSM instruments, Buchs, Switzerland). The load was 1 N for all experiments but with various impact height, thereby various effect of impact, namely 0.2 mm, 1.0 mm, 2.0 mm, and 3.3 mm. The pin was lifted at a frequency of 1 Hz and in contact with the sample for a period close to 0.5 s. The pin hit the sample at a near but different place each time, forming a round wear track eventually. The radius was 8 mm, and the rotational speed was 2 Hz (corresponding linear speed is 0.1 mm/s) for each 2.75-h experiment. Three liquids were used to provide different corrosion condition, namely non-corrosive ethanol, corrosive deionized water, and severely corrosive seawater (simulated by a 3.5% NaCl solution). Each combination of impact height and liquid was repeated three times to obtain the variability of the experiment. The counterpart used in this study was an aluminum oxide ball with a diameter of 6 mm. After each experiment, the ball was either rotated or replaced to obtain fresh contact between the sample and the ball.

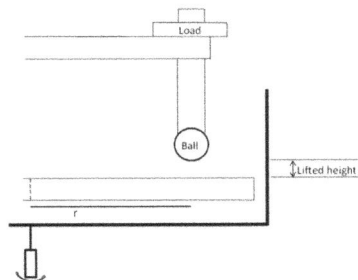

**Figure 1.** The diagram of the modified pin on disc. A lifted distance exists between tha ball and the sample surface. The distance is well controlled and measured by the software of pin on disc.

### 2.3. Characterization

After experiments, the wear track profiles of the samples were measured by white light interferometry (Bruker Contour GT-X, Bruker, Leiderdorp, The Netherlands). Specifically, four different places of each cross section of the wear track were measured, and the wear rate was calculated by multiplying the area of the cross section by the perimeter of the wear track, divided by the sliding distance [16].

### 3. Results

#### 3.1. Wear Rate Comparison

Wear rate, overall, increased with increasing impact height from 0.2 mm to 3.3 mm, as shown in Figure 2a. For the case of 0.2 mm, the wear rate in deionized water was the largest, followed by wear in ethanol. In seawater, the wear rate was the smallest. This order is the same as was obtained in sliding wear in these liquids [14]. For higher impact heights, the wear rate difference among three liquids decreases. At 1.0 mm impact height, the wear rate in the three liquids is identical within the error of measurement. For higher impact heights the differences are even smaller. The net material loss, shown in Figure 2b, shows the same behavior, except that the wear rate increase with the increasing impact height is much smaller.

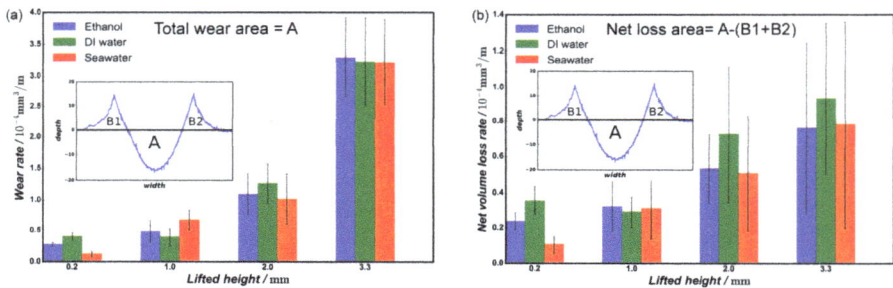

**Figure 2.** (**a**) Wear rate comparison at various impact heights; (**b**) Net volume loss comparison at various impact height. Ethanol, deionized (DI) water and seawater were used to vary corrosivity. The error bar stands for one standard deviation of three repeated results. The embedded small figure represents the cross section of the round wear track.

#### 3.2. Wear Track Analysis

The wear track comparison at various impact height in three liquids is illustrated in Figure 3. As can be seen, for each liquid, with increasing impact height, the depth of the track increases as well, and when the impact height reaches 3.3 mm, the wear depth seems to be the same among all liquids, well corresponding to the wear rate comparison (Figure 2). The total volume loss due to wear has two components: material removal and deformation. The deformation part accounts for a large portion of the total wear when the impact height is beyond 1 mm. As shown in Figure 4, for all three liquids, the ratio of deformation to total wear is less than 20% for the impact height of 0.2 mm, by comparison, it reaches nearly 80% when the impact height is 3.3 mm. For 1 mm and 2 mm, the ratios are comparable. In seawater, the ratio change is the most dramatic from less than 20% to roughly 60%, when the impact height increases from 0.2 mm to 1 mm. Additionally, the wear track, overall, is smooth except for the situation where the impact height is the least, 0.2 mm, which shows a level of roughness, similar to the situation in pure sliding [14].

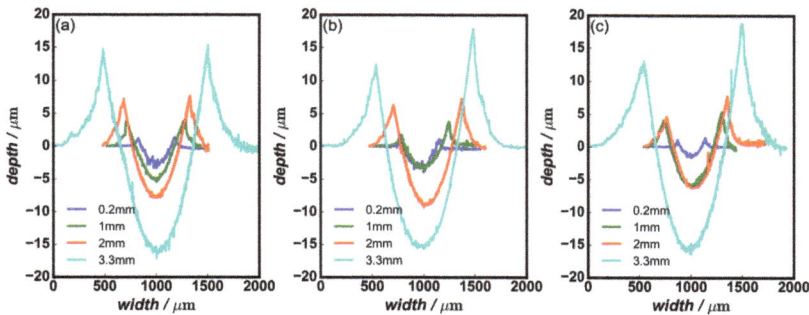

**Figure 3.** Wear track comparison at various impact heights in three liquids: (**a**) in ethanol; (**b**) in DI water; (**c**) in sea water. The average of three repetitive results was used for each situation. The wear track was moved to the center in order to be easily compared.

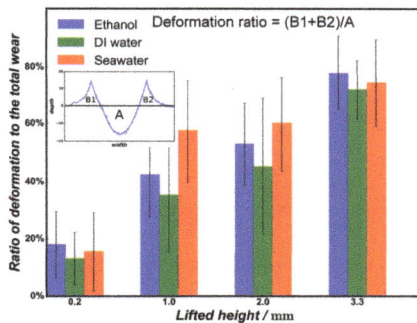

**Figure 4.** The ratio of plastic deformation to the total wear at various impact heights. Ethanol, DI water and seawater were used to vary corrosivity. The error bar stands for one standard deviation of three repeated results. The embedded small figure represents the cross section of the round wear track.

## 4. Discussion

In slurry transport, the pipes wear due to the combination of sliding, impact and corrosion [1]. The interaction among those factors determines the final wear. Typically, the interaction leads to a positive synergistic effect and results in higher wear rate. However, research has shown the beneficial effect of the interaction between corrosion and abrasion, where corrosion is able to modify the sample surface by dissolving the soft phase and leaving the hard phase protruding, and subsequently, abrasion redistributes the hard phase, increasing the wear resistance [14]. However, in real slurry transport situations, no noticeable wear difference between corrosive and non-corrosive mediums was observed and reported. Although the scale of the research in a lab is different from in real situation, the mechanisms should keep the same. In this study, impact shows a huge influence on the wear rate. At 0.2 mm, the wear rate in ethanol, DI water, and seawater shows a huge difference, and in seawater, the wear rate is the smallest. However, when the impact height increases to 1 mm, the wear rate difference is within the experimental uncertainty. From 1 mm impact height, the influence of corrosion is not able to dominate. At 0.2 mm, the impact is small, and the wear result is similar to the result found in pure sliding. In pure sliding, the modified surface is able to increase the wear resistance of the sample, showing a beneficial effect. From low impact to high impact, the wear mechanism changes from material removal to mainly plastic deformation, as shown in Figure 4. When deformation becomes the determining factor of wear, surface modification, which governs the wear mechanism in pure sliding wear, does not make a difference. The deformation also results in a rather smooth wear track, which is not typical in a corrosive environment. The smooth wear track implies that the influence of corrosion becomes a minor influencing factor.

## 5. Conclusions

In this study, the effect of impact on the wear of a structural steel S235 was studied in various liquids. When impact is small, corrosion plays a dominant role so that the sample in seawater wears the least due to the beneficial interaction with abrasion. However, when impact is large, the wear rates among three liquids do not show a noticeable difference, because the wear mechanism changes from material removal to mainly plastic deformation. The influence of corrosion becomes a minor influencing factor when subject to impact.

**Supplementary Materials:** The followings are available online at http://www.mdpi.com/2079-6412/7/12/237/s1, Figure S1: Estimated hitting speed of the pin: 0.2 mm; 1 mm; 2 mm; 3.3 mm; Figure S2: The micrograph of worn surfaces imaged with Scanning Electron Microscope; Table S1: Estimated hitting speed and angle of the pin on the sample for various lifted heights.

**Acknowledgments:** This research was carried out under project number M33.3.11427b in the framework of the Research Program of the Materials Innovation Institute M2i (www.m2i.nl).

**Author Contributions:** Yueting Liu and G.C.A.M. Janssen conceived the idea and designed the experiments. Yueting Liu performed the experiments and wrote the paper.

**Conflicts of Interest:** The authors declare no conflicts of interest.

## References

1.  Wilson, K.C.; Addie, G.R.; Sellgren, A.; Clift, R. *Slurry Transport Using Centrifugal Pumps*, 3rd ed.; Springer: Boston, MA, USA, 2006.
2.  Roco, M.C.; Addie, G.R. Erosion wear in slurry pumps and pipes. *Powder Technol.* **1987**, *50*, 35–46. [CrossRef]
3.  Jones, M.; Llewellyn, R.J. Erosion-corrosion assessment of materials for use in the resources industry. *Wear* **2009**, *267*, 2003–2009. [CrossRef]
4.  Shivamurthy, R.C.; Kamaraj, M.; Nagarajan, R.; Shariff, S.M.; Padmanabham, G. Influence of microstructure on slurry erosive wear characteristics of laser surface alloyed 13Cr-4Ni steel. *Wear* **2009**, *267*, 204–212. [CrossRef]
5.  Ramesh, C.S.; Keshavamurthy, R.; Channabasappa, B.H.; Pramod, S. Influence of heat treatment on slurry erosive wear resistance of Al6061 alloy. *Mater. Des.* **2009**, *30*, 3713–3722. [CrossRef]
6.  Bross, S.; Addie, G. Prediction of impeller nose wear behavior in centrifugal slurry pumps. *Exp. Therm. Fluid Sci.* **2002**, *26*, 841–849. [CrossRef]
7.  Dube, N.M.; Dube, A.; Veeregowda, D.H.; Iyer, S.B. Experimental technique to analyze the slurry erosion wear due to turbulence. *Wear* **2009**, *267*, 259–263. [CrossRef]
8.  Tupper, G.B.; Govender, I.; Mainza, A.N.; Plint, N. A mechanistic model for slurry transport in tumbling mills. *Miner. Eng.* **2013**, *43–44*, 102–104. [CrossRef]
9.  Ojala, N.; Valtonen, K.; Antikainen, A.; Kemppainen, A.; Minkkinen, J.; Oja, O.; Kuokkala, V.-T. Wear performance of quenched wear resistant steels in abrasive slurry erosion. *Wear* **2016**, *354–355*, 21–31. [CrossRef]
10. Rajahram, S.S.; Harvey, T.J.; Wood, R.J.K. Erosion-corrosion resistance of engineering materials in various test conditions. *Wear* **2009**, *267*, 244–254. [CrossRef]
11. Truscott, G.F. A literature survey on abrasive wear in hydraulic machinery. *Wear* **1972**, *20*, 29–50. [CrossRef]
12. Deng, T.; Bingley, M.S.; Bradley, M.S.A.; de Silva, S.R. A comparison of the gas-blast and centrifugal-accelerator erosion testers: The influence of particle dynamics. *Wear* **2008**, *265*, 945–955. [CrossRef]
13. Kotzur, B.A.; Berry, R.J.; Bradley, M.S.; Farnish, R.J. Quantifying the influence of secondary impacts within centrifugal impact testers. In Proceedings of the 12th International Conference on Bulk Materials Storage, Handling and Transportation (ICBMH 2016), Darwin, Austrilia, 11–14 July 2016; pp. 373–382.

14. Liu, Y.; Mol, J.M.C.; Janssen, G.C.A.M. Corrosion reduces wet abrasive wear of structural steel. *Scr. Mater.* **2015**, *107*, 92–95. [CrossRef]
15. *European Standard EN 10025-2:2004 Hot Rolled Products of Structural Steels—Part 2: Technical Delivery Conditions for Non-Alloy Structural Steels*; British Standard Institution: London, UK, 2004.
16. Rabinowicz, E.; Tanner, R.I. Friction and wear of materials. *J. Appl. Mech.* **1966**, *33*, 479. [CrossRef]

*coatings*

MDPI

*Article*

# Effects of Laser Processing Parameters on Texturized Layer Development and Surface Features of Ti6Al4V Alloy Samples

Juan Manuel Vázquez Martínez [1,*], Jorge Salguero Gómez [1], Moises Batista Ponce [1] and Francisco Javier Botana Pedemonte [2]

[1] Department of Mechanical Engineering & Industrial Design, Faculty of Engineering, University of Cadiz, Av. Universidad de Cadiz 10, E-11519 Puerto Real-Cadiz, Spain; jorge.salguero@uca.es (J.S.G.); moises.batista@uca.es (M.B.P.)

[2] Department of Materials Science and Metallurgic Engineering and Inorganic Chemistry, Faculty of Engineering, University of Cadiz, Av. Universidad de Cadiz 10, E-11519 Puerto Real-Cadiz, Spain; javier.botana@uca.es

* Correspondence: juanmanuel.vazquez@uca.es

Academic Editor: Alicia Esther Ares
Received: 30 November 2017; Accepted: 20 December 2017; Published: 22 December 2017

**Abstract:** Surface engineering is widely used in different areas, such as the aerospace industry or the biomechanical and medical fields. Specifically, laser surface modification techniques may obtain specific surface finishes for special applications. In texturing laser procedures, the control of processing parameters has a great influence on the geometry and characteristics of the treated area. When these processes are carried out on titanium alloys, thin oxide layers are usually developed on the irradiated surface, formed through the thermochemical combination of vaporized material with atmospheric oxygen in the air. In thermal oxidation treatments of Ti6Al4V, the highest concentration of oxides is mainly composed by rutile ($TiO_2$), producing surface property modifications such as hardness, among others. In this research, a thermochemical oxidation of Ti6Al4V alloy has been performed through laser texturing, using laser scanning speed ($V_s$) and pulse rate ($f$) as process control variables, and its influence on the beam absorption capacity of the modified layer have been analyzed. Combined evaluations of microgeometrical features and mechanical properties, such as hardness, verified that, by means of laser texturing treatments, the ability to generate specific topographies and increase the initial hardness of the alloy is obtained. The most advantageous results for the increase of hardness by thermochemical oxidation have been detected in low scan speeds of laser beam treatments, resulting in an increase of approximately 270% using a scanning speed of 10 mm/s. On the other hand, a dependence between roughness values, in terms of $R_a$ and $R_z$, and the energy density of pulse ($E_d$) has been observed, showing higher values of roughness for a 17.68 J/cm² energy density of pulse.

**Keywords:** surface engineering; laser texturing; thermal oxidation; Ti6Al4V; surface characterization

---

## 1. Introduction

Titanium alloys are one of the most commonly used materials in strategic fields such as the aerospace and biomechanical industry. This fact is mainly due to the ratio between physicochemical properties and weight, and an excellent biocompatibility with organic environments [1–3]. Also, current advances in biotechnology and biomechanics induce the development of new materials and surface treatments that aim to increase wear and corrosion strength, or improve features, which allows the promotion of the biointegration of materials [4–8]. Laser processing techniques imply the ability to perform changes on surface material without direct contact [9–13]. This feature results in a relatively clean and precise process, due to an excellent control over the heat absorption on the affected area, and the possibility of focusing the laser beam on small areas of the surface. The exposure of Ti6Al4V alloy to radiation treatments by laser techniques in an air atmosphere causes the combination of the external surface layer with the oxygen present in ambient air, promoted by the temperature increase generated in the radiation process [14–17]. Such a combination results in the development of a thin protective layer, mainly composed of titanium oxides, with rutile ($TiO_2$) being the predominant form [15–17]. A range of several tonalities is shown according to variables such as the thickness and microstructure of the oxidation layer [15–17]. With the aim of ensuring the quality of results, metrological procedures are required to evaluate the dimensions of the grooves created by surface laser texturing [18].

Through variations of conditions of laser treatment, several features, such as the surface finish or the hardness, can be tuned for very specific material uses [19–21]. Based on laser surface treatments, a large number of studies are focused on improving the tribological properties of titanium alloys, induced by limitations and poor material behavior in friction sliding conditions [22,23].

In the present work, laser texturing has been performed on Ti6Al4V alloy surfaces, combining energy density on pulse ($E_d$) and speed scans of beam ($V_s$) values. In this way, the influence of intrinsic parameters of the laser processing setup on morphology, characteristics and properties of treated surfaces has been analyzed.

## 2. Materials and Methods

### 2.1. Laser Texturing

Surface texturing aims to develop outer layers of reduced thickness with specific microstructural properties, decreasing the heated irradiated area. In this study, the laser texturing of a Ti6Al4V alloy has been carried out using an ytterbium fiber infrared pulsed laser system Rofin EasyMark F20 (ROFIN-SINAR Technologies Inc., Plymouth, MI, USA) with $1070 \pm 5$ nm wavelength, 60 μm spot diameter and 100 ns pulse width (according to the manufacturer). Surface treatments have been developed under ambient air conditions in order to generate the oxidation of the surface alloy. A bidirectional traces layout has been performed without overlapping throughout the titanium samples, resulting in parallel textured lines with approximately 0.12 mm between neighboring laser traces. With the purpose of obtaining a higher uniformity in the radiation absorption capacity of the alloy's surface, an adaptation process has been realized by grinding the initial surface finish to 1200 grits (particles per square inch). Subsequently, through parameter combinations of pulse energy ($E_t$), fluence or energy density of pulse ($E_d$) and scanning speed ($V_s$), 24 different types of specimens were obtained, keeping a constant laser power of 10 W. With these combinations, a sufficiently broad range of treatments are analyzed to study the behavior of the alloy under different radiation stages—see Table 1 below.

**Table 1.** Laser processing parameters.

| Parameter | $f_1$ | | | $f_2$ | | | $f_3$ | |
|---|---|---|---|---|---|---|---|---|
| $f$ (Hz) | 20,000 | | | 50,000 | | | 80,000 | |
| $E_t$ (mJ) | 0.500 | | | 0.200 | | | 0.125 | |
| $E_d$ (J/cm$^2$) | 17.68 | | | 7.07 | | | 4.42 | |
| Parameter | $V_1$ | $V_2$ | $V_3$ | $V_4$ | $V_5$ | $V_6$ | $V_7$ | $V_8$ |
| $V_s$ (mm/s) | 10 | 20 | 40 | 80 | 100 | 150 | 200 | 250 |

### 2.2. Evaluation of Laser Textured Specimens

In order to assess the effects caused by laser texturing on the surface of titanium alloy specimens, several procedures have been developed, including visual and microscopy inspection, microstructural and compositional analysis, measurement of surface finish, depth evaluation of molten material and hardness variation induced by oxidation.

The main purpose of visual and microscopy inspection is the detection of surface variations through chromatic changes and microstructural modifications. In order to obtain higher contrast and enhance the detection of different phases and grain sizes of the material surface, a metallographic setup and subsequent chemical attack by Kroll's reagent technique on transverse sections of the irradiated surface have been performed. After sampling the metallographic setup, energy dispersive spectrometry (EDS) analysis has been conducted to evaluate the indications of oxides on the substrate material.

In this work, surface finish measurements have been applied to textured surfaces by using a roughness measurement device: model Mahr Perthometer Concept PGK120 (Mahr technology, Göttingen, Germany). With the aim of characterizing the surface integrity from the microgeometrical point of view, linear profile measurements have been carried out along the entire surface, perpendicularly to the displacement of the laser beam, in order to quantify the average roughness parameter, $R_a$.

By optical and scanning electron microscopy (SEM), some features in the shape of the radiation grooves, undetectable at surface level, have been found. To do this, a metallographic setup was required to inspect the internal section in order to obtain a deeper description of the heat-affected area of laser processing. Subsequently, a depth measurement of the carved grooves, quantifying the thickness of the surface layer affected by the process, has also been performed.

Finally, hardness measurements have been carried out on cross sections of laser-treated samples in a near-surface layer of material, where the thermal affected area and higher oxide concentration gradients are located. To perform this measurement, a Shimadzu (RTY) microdurometer (Kyoto, Japan) has been used with a test load of 245.2 mN (HV 0.025) for a time of 10 s. In order to analyze the goodness of measurements, a series of measurements have been made on untreated Ti6Al4V alloy, obtaining less than a 3% variation from the nominal value of material hardness.

## 3. Results and Discussion

### 3.1. Laser Textured Specimens: Chromatic Variability

During laser texturing, particles located in the outer layers of the Ti6Al4V alloy react with the oxygen present in the atmosphere, causing oxidation of the material surface and forming a thin layer of titanium oxide ($Ti_yO_x$). This behavior is mainly caused by the high temperatures produced during laser processing and may also be observed during hard conventional machining processes such as dry drilling in the Ti6Al4V alloy [24]. Regarding this behavior, the oxidation layer is usually composed of various types of oxide, predominantly rutile $TiO_2$ form. Due to the high reactivity between titanium and oxygen particles present in ambient air, a wide range of oxidation states can be developed, being characterized by their color tonality. In this way, through the variation of laser processing parameters (energy density of pulse, scanning speed) associated with the presence of oxidation layers, different treatment intensities may be achieved, obtaining chromatic variation, different thickness and characteristics—see Figure 1.

In the same way as in anodizing processes, the thickness of the modified layer subjected to oxidation is shown as one of the main parameters that affects the chromatic variation of the material [25]. In this aspect, in accordance with Perez del Pino et al. [15,16], it has been detected that, for treatments that involve higher thicknesses of modified layers corresponding to lower $E_d$ and $V_s$, there is a tendency towards bluish and grayish to dark brown color tonalities. On the other hand, by increasing $V_s$ and/or reducing $E_d$, developing modified layers with lower thickness, light brown to golden color tonalities textures are generated.

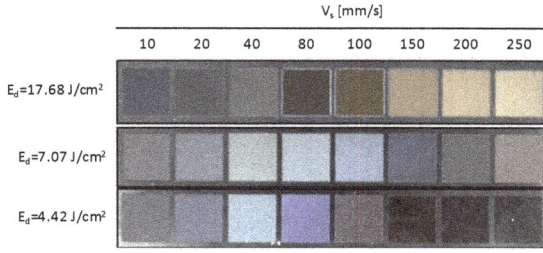

**Figure 1.** Tonality variations on thermal Ti6Al4V oxidation layers.

## 3.2. Laser Textured Specimens: Microstructure and Composition

The grade 5 titanium alloy (ASTM) used in this study shows an equiaxed or mill-annealed microstructure type, formed by fine grains of α- (lighter) and β- (darker) phases. The surface fusion procedure by laser may induce the generation of microstructures which differ from the initial Ti6Al4V alloy, allowing us also to determine the existence of a thermally affected area [26]. It is observed that the zone affected by direct incidence of laser pulses presents variations in the initial composition, surface finish and texture, enabling us to detect alterations in substrate microstructure and oxidation rates of the alloy—see Figure 2.

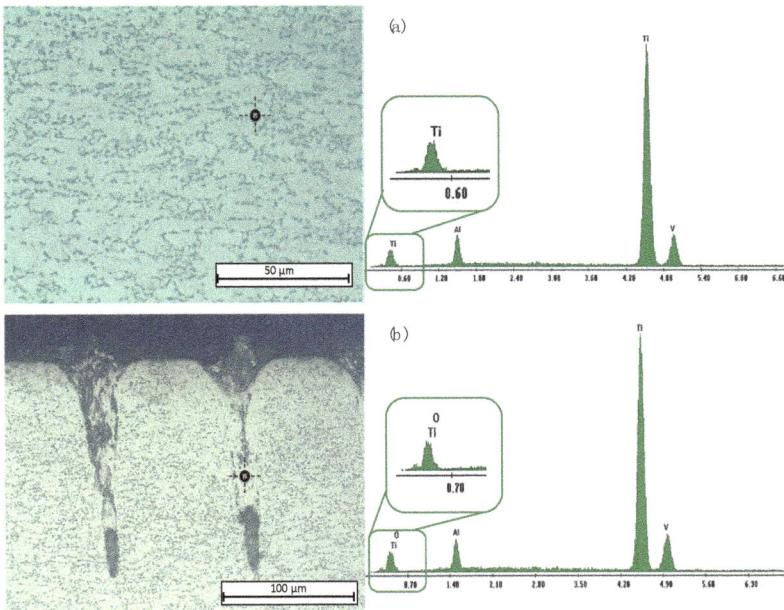

**Figure 2.** (a) Microstructure and EDS analysis of Ti6Al4V surface; (b) Cross section and EDS analysis of heat affected zone of laser incidence for 17.68 J/cm$^2$; 10 mm/s texturing conditions.

## 3.3. Laser Textured Specimens: Surface Finish (Roughness)

Laser processing parameters have a great influence on the resulting roughness in texturing processes. Therefore, in some cases, a precise control of them may induce the generation of specific microgeometry topographies [17].

Starting from polished Ti6Al4V samples with average values of arithmetic average roughness $R_a = 0.049 \pm 0.005$ µm and maximum height of roughness profile $R_z = 0.405 \pm 0.070$ µm, the surface finish of the treatments performed indicates that variations in laser texturing parameters have a great influence on the absorption of laser radiation, obtaining a significant variability of roughness values. Moreover, different shapes can be observed in the profiles obtained for different combinations of pulse rate and scanning speed—see Figure 3.

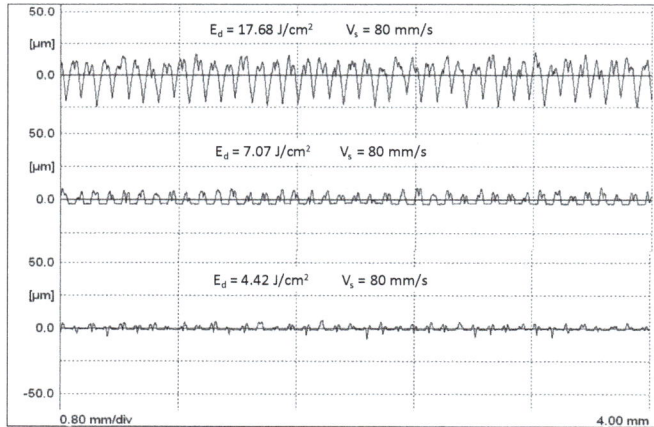

**Figure 3.** Roughness profiles as a function of energy density of pulse and scan speed of the beam.

Specific combinations of fluence and scan speed result in surface textures with different natures, mainly based on the cooling process and material melted rate. However, it is noteworthy that the energy density of the pulse is shown as the greatest influential parameter on the development of roughness—see Figure 4a,b. In this aspect, an increase in the pulse rate results in the melting of a higher material thickness, producing a slower solidification process and smoothing the topography of the modified surface. As a consequence of this, more aggressive treatments related to higher energy density absorbed by the surface can result in more volume of melted alloy, increasing the depth and height of the asperities produced by the texturing process.

Analyzing the topography from laser texturing tracks, a wide range of grooves with varying dimensions and geometry may be detected. With the aim of obtaining accurate measurements and an overview of the removed material grooves, a metallographic setup of cross sections of the textured samples has been conducted.

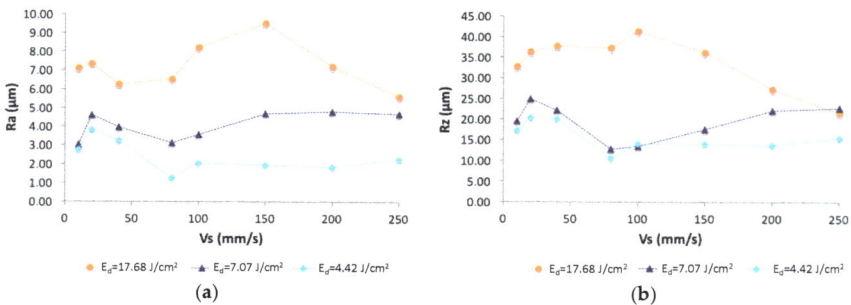

**Figure 4.** Surface finish (**a**) $R_a$ and (**b**) $R_z$ behavior with different laser processing parameters.

*3.4. Laser Textured Specimens: Laser Grooves Depth*

In order to obtain accurate information about the surface behavior after the radiation process and to evaluate possible changes in the microstructure of the alloy sections unaffected by thermal oxidation, metallographic preparations of treated samples have been carried out. Thus, through an acid etching procedure, grain boundaries that form the $\alpha$- and $\beta$-phases of the alloy Ti6Al4V can be clearly revealed.

Due to variations in the frequency and scan speed of the laser parameters, several types of texture and grooves can be observed, influencing the geometry and reaching depth [27], as can be seen in Figure 5. This analysis allows us to perform approaches to the behavior of the material after it has been vaporized by laser irradiation, favoring the evaluation of the heat-affected zone.

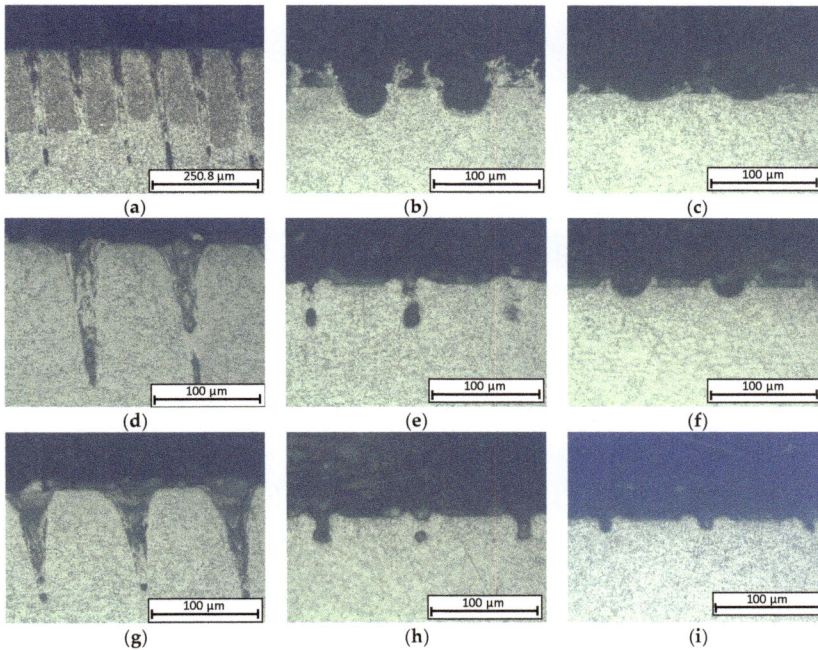

**Figure 5.** Different morphology of grooves due to laser absorption features. (**a**) 17.68 J/cm$^2$, 10 mm/s; (**b**) 17.68 J/cm$^2$, 80 mm/s; (**c**) 17.68 J/cm$^2$, 250 mm/s; (**d**) 7.07 J/cm$^2$, 10 mm/s; (**e**) 7.07 J/cm$^2$, 80 mm/s; (**f**) 7.07 J/cm$^2$, 250 mm/s; (**g**) 4.42 J/cm$^2$, 10 mm/s; (**h**) 4.42 J/cm$^2$, 80 mm/s; (**i**) 4.42 J/cm$^2$, 250 mm/s.

The depth of grooves produced by laser beam incidence shows direct relationships with the processing parameters. Laser scan speed is presented as the most influential parameter in the track depth value generated by radiation. However, the increase of pulse energy coincident with lower frequency values bound to a greater beam exposure time on a single point, resulting in a significant increase of thickness in the affected layer of the alloy—see Figure 6.

A maximum depth value of approximately 300 μm is observed for the highest values of the fluence and scan speed of the beam parameters, showing similar behaviors for the three values of energy density of pulse studied, only decreasing the depth for the lower values. In this aspect, when the scan speed increases, an absorption of a lower density of pulses takes place on the same section of the surface, tending to decrease the incidence depth of the laser beam, developing softer topographies as expected.

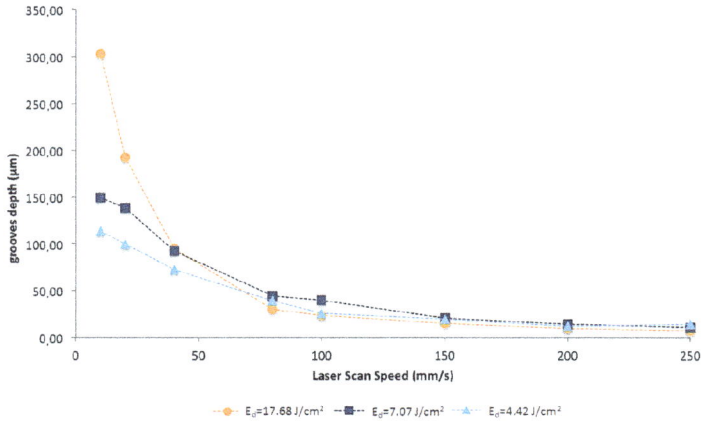

**Figure 6.** Grooves depth as a function of laser processing parameters.

### 3.5. Laser Textured Specimens: Hardness Variation

Microstructural modification and thermal oxidation of the alloy is usually associated with a hardness variation of the material, as such procedures are used, in most cases, in hardening processes. After hardness measurement located on the cross sections of the texturing grooves, it has been observed that the hardness of the material increases considerably depending on the proximity to the laser irradiated section, obtaining values that can triple the initial value of the base alloy Ti6Al4V—see Figure 7.

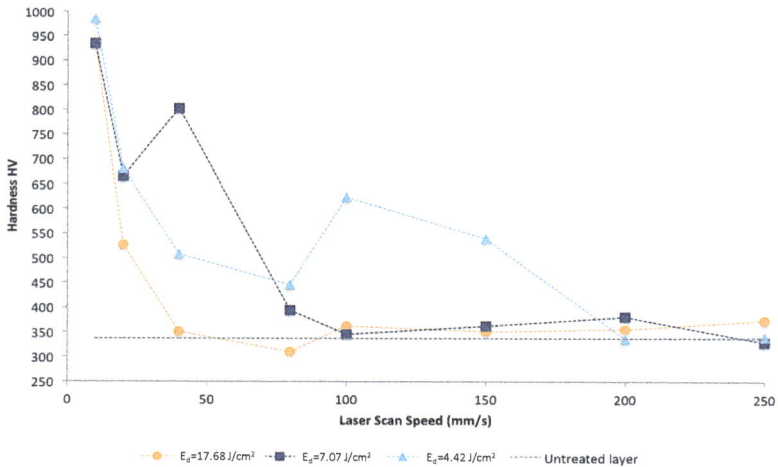

**Figure 7.** Hardness as a function of laser processing parameters.

As previously indicated, a significant influence of laser parameters can be observed on hardness under thermal oxidation conditions. The proximity to the material surface, and therefore to the remelted layer, results in a hardness increase, expressed with dimensions of Vickers test traces. This fact is due to the existence of microstructural changes of the alloy and oxidation reactions.

As can be appreciated in Figure 7, there is a clear dependence between hardness values and the scan speed of the laser beam, increasing significantly with lower speeds, which generates samples with higher thicknesses in the irradiated layers.

## 4. Conclusions

Variations of laser processing parameters may induce the formation of modified material layers with different natures and properties. The energy density of the pulse (fluence) ($J/cm^2$) and scanning speed of the beam are the main variables that govern texturing processes.

The different treatments obtained through setup parameter combinations allow the obtaining of samples with widely varying features, according to the thickness of the treated layer and incidence depth of laser beam. This diversity of samples shows a range of color tonalities from soft gold to medium gray and blue.

Laser surface texturing processes greatly affect roughness profiles, with the energy density of pulse ($E_d$) being the most relevant parameter in controlling the properties of surface finish. Furthermore, the intensity of treatments, in terms of the incidence absorption of the beam, result in the modification of microstructure and composition (with addition of oxide concentrations) of the outer layers of the alloy. Microstructural changes and oxidation layer generation provide a significant increase in material hardness, reaching Vickers hardness values close to 1000 HV on high intensity treatments.

**Acknowledgments:** This work has received financial support by the Spanish Government and the Andalusian Government (PAIDI).

**Author Contributions:** Juan Manuel Vázquez Martínez, Jorge Salguero Gómez and Francisco Javier Botana Pedemonte conceived and designed the experiments; Juan Manuel Vázquez Martínez and Moises Batista Ponce performed the experiments; Juan Manuel Vázquez Martínez, Francisco Javier Botana Pedemonte and Jorge Salguero Gómez analyzed the data; Juan Manuel Vázquez Martínez wrote the paper.

**Conflicts of Interest:** The authors declare no conflict of interest.

## References

1.  Ehtemam-Haghighi, S.; Prashanth, K.G.; Attar, H.; Chaubey, A.K.; Cao, G.H.; Zhang, L.C. Evaluation of mechanical and wear properties of Ti-xNb-7Fe alloys designed for biomedical applications. *Mater. Des.* **2016**, *111*, 592–599. [CrossRef]
2.  Okulov, I.V.; Volegov, A.S.; Attar, H.; Bönisch, M.; Ehtemam-Haghighi, S.; Calin, M.; Eckert, J. Composition optimization of low modulus and high-strength TiNb-based alloys for biomedical applications. *J. Mech. Behav. Biomed. Mater.* **2017**, *65*, 866–871. [CrossRef] [PubMed]
3.  Okulov, I.V.; Wendrock, H.; Volegov, A.S.; Attar, H.; Kühn, U.; Skrotzki, W.; Eckert, J. High strength beta titanium alloys: New design approach. *Mater. Sci. Eng. A* **2015**, *628*, 297–302. [CrossRef]
4.  Le Guéhennec, L.; Souedidan, A.; Layrolle, P.; Amouriq, Y. Surface treatments of titanium dental implants for rapid osseointegration. *Dent. Mater.* **2007**, *23*, 844–854. [CrossRef] [PubMed]
5.  Ponsonnet, L.; Reybier, K.; Jaffrezic, N.; Comte, V.; Lagneau, C.; Lissac, M.; Martelet, C. Relationship between surface properties (roughness, wettability) of titanium and titanium alloys and cell behavior. *Mater. Sci. Eng. C* **2003**, *23*, 551–560. [CrossRef]
6.  Rosales-Leal, J.I.; Rodriguez-Valverde, M.A.; Mazzaglia, G.; Ramon-Torregrosa, P.J.; Diaz-Rodriguez, L.; Garcia-Martinez, O.; Vallecillo-Capilla, M.; Ruiz, C.; Cabrerizo-Vilchez, M.A. Effects of roughness, wettability and morphology of engineered titanium surfaces on osteoblast-like cell adhesion. *Colloid Surf. A Physicochem. Eng. Asp.* **2010**, *365*, 222–229. [CrossRef]
7.  Le Guehennec, L.; Lopez-Heredia, M.A.; Enkel, B.; Weiss, P.; Amouriq, Y.; Layrolle, P. Osteoblastic cell behavior on different titanium implant surfaces. *Acta Biomater.* **2008**, *4*, 535–543. [CrossRef] [PubMed]
8.  Feng, B.; Weng, J.; Yang, B.C.; Qu, S.X.; Zhang, X.D. Characterization of surface oxide films on titanium and adhesion of osteoblast. *Biomaterials* **2003**, *24*, 4663–4670. [CrossRef]
9.  Amaya-Vazquez, M.R.; Sanchez-Amaya, J.M.; Boukha, Z.; Botana, F.J. Microstructure, microhardness and corrosion resistance of remelted TiG$_2$ and Ti6Al4V by a high power diode laser. *Corros. Sci.* **2012**, *56*, 36–48. [CrossRef]

10. Lednev, V.N.; Pershin, S.M.; Ionin, A.A.; Kudryashov, S.I.; Makarov, S.V.; Ligachev, A.E.; Rudenko, A.A.; Chmelnitsky, R.A.; Bunkin, A.F. Laser ablation of polished and nanostructured titanium surfaces by nanosecond laser pulses. *Spectrochim. Acta Part B* **2013**, *88*, 15–19. [CrossRef]
11. Weng, F.; Chuanzhong, C.; Yu, H. Research status of laser cladding on titanium and its alloys: A review. *Mater. Des.* **2014**, *58*, 412–425. [CrossRef]
12. Attar, H.; Ehtemam-Haghighi, S.; Kent, D.; Okulov, I.V.; Wendrock, H.; Bönisch, M.; Volegov, A.S.; Calin, M.; Eckert, J.; Dargusch, M.S. Nanoindentation and wear properties of Ti and Ti-TiB composite materials produced by selective laser melting. *Mater. Sci. Eng. A* **2017**, *688*, 20–26. [CrossRef]
13. Attar, H.; Ehtemam-Haghighi, S.; Kent, D.; Wu, X.; Dargusch, M.S. Comparative study of commercially pure titanium produced by laser engineered net shaping, selective laser melting and casting processes. *Mater. Sci. Eng. A* **2017**, *705*, 385–393. [CrossRef]
14. Wen, M.; Wen, C.; Hodgson, P.; Li, Y. Thermal oxidation behavior of bulk titanium with nanocrystalline surface layer. *Corros. Sci.* **2012**, *59*, 352–359. [CrossRef]
15. Pérez del Pino, A.; Serra, P.; Morenza, J.L. Coloring of titanium by pulsed laser processing in air. *Thin Solid Films* **2002**, *415*, 201–205. [CrossRef]
16. Pérez del Pino, A.; Fernandez-Pradas, J.M.; Serra, P.; Morenza, J.L. Coloring of titanium through laser oxidation: Comparative study with anodizing. *Surf. Coat. Technol.* **2004**, *187*, 106–112. [CrossRef]
17. Adams, D.P.; Murphy, R.D.; Saiz, D.J.; Hirschfeld, D.A.; Rodriguez, M.A.; Kotula, P.G.; Jared, B.H. Nanosecond pulsed laser irradiation of titanium: Oxide growth and effects on underlying metal. *Surf. Coat. Technol.* **2014**, *248*, 38–45. [CrossRef]
18. Vazquez-Martinez, J.M.; Salguero, J.; Botana, F.J.; Contreras, J.P.; Fernandez-Vidal, S.R.; Marcos, M. Metrological evaluation of the tribological behavior of laser surface treated Ti6Al4V alloy. *Procedia Eng.* **2013**, *63*, 752–760. [CrossRef]
19. Ukar, E.; Lamikiz, A.; Martinez, S.; Tabernero, I.; López de Lacalle, L.N. Roughness prediction on laser polished surfaces. *J. Mater. Process. Technol.* **2012**, *212*, 1305–1313. [CrossRef]
20. Lavisse, L.; Jouvard, J.M.; Imhoff, L.; Heintz, O.; Korntheuer, J.; Langlade, C.; Bourgeois, S.; Marco de Lucas, M.C. Pulsed laser growth and characterization of thin films on titanium substrates. *Appl. Surf. Sci.* **2007**, *253*, 8226–8230. [CrossRef]
21. Mahamood, R.M.; Akinlabi, E.T.; Shukla, M.; Pityana, S. Scanning velocity influence on microstructure, microhardness and wear resistance of laser deposited Ti6Al4V/TIC composite. *Mater. Des.* **2013**, *50*, 656–666. [CrossRef]
22. Vazquez-Martinez, J.M.; Salguero, J.; Botana, F.J.; Gomez-Parra, A.; Fernandez-Vidal, S.R.; Marcos, M. Tribological wear analysis of laser surface treated Ti6Al4V based on volume lost evaluation. *Key Eng. Mater.* **2014**, *615*, 82–87. [CrossRef]
23. Dongqin, H.; Zheng, S.; Pu, J.; Zhang, G.; Hu, L. Improving tribological properties of titanium alloys by combining laser surface texturing and diamond-like carbon film. *Tribol. Int.* **2015**, *82*, 20–27. [CrossRef]
24. Salguero, J.; Batista, M.; Sanchez, J.A.; Marcos, M. An XPS study of the stratified Built-Up layers developed onto the tool surface in the dry drilling of Ti alloys. *Adv. Mater. Res.* **2011**, *223*, 564–572. [CrossRef]
25. Vera, M.L.; Avalos, M.C.; Rosenberger, M.R.; Bolmaro, R.E.; Schvezov, C.E.; Ares, A.E. Evaluation of the influence of texture and microstructure of titanium substrates on TiO$_2$ anodic coatings at 60 V. *Mater. Charact.* **2017**, *131*, 348–358. [CrossRef]
26. Vrancken, B.; Thijs, L.; Kruth, J.P.; van Humbeeck, J. Heat treatment of Ti6Al4V produced by selective laser melting: Microstructure and mechanical properties. *J. Alloys Compd.* **2012**, *541*, 177–185. [CrossRef]
27. Fasai, A.Y.; Mwenifumbo, S.; Rahbar, N.; Chen, J.; Li, M.; Beye, A.C. Nano-second UV laser processed micro-grooves on Ti6Al4V for biomedical applications. *Mater. Sci. Eng. C* **2009**, *29*, 5–13. [CrossRef]

*coatings*

MDPI

*Article*

# Effect of Tip Shape of Frictional Stir Burnishing Tool on Processed Layer's Hardness, Residual Stress and Surface Roughness

**Yoshimasa Takada [1] and Hiroyuki Sasahara [2,\*]**

[1]   Industrial Division, Nikkiso Co., Ltd., 2-16-2 Noguchicho, Higashimurayama-shi, Tokyo 189-8520, Japan;
     y.takada@nikkiso.co.jp
[2]   Department of Mechanical Systems Engineering, Faculty of Engineering, Tokyo University of Agriculture
     and Technology, 2-24-16 Nakacho, Koganei-shi, Tokyo 184-8588, Japan
*    Correspondence: sasahara@cc.tuat.ac.jp; Tel.: +81-42-388-7240

Received: 30 November 2017; Accepted: 9 January 2018; Published: 11 January 2018

**Abstract:** Friction stir burnishing (FSB) is a surface-enhancement method used after machining, without the need for an additional device. The FSB process is applied on a machine that uses rotation tools (e.g., machining center or multi-tasking machine). Therefore, the FSB process can be applied immediately after the cutting process using the same machine tool. Here, we apply the FSB to the shaft materials of 0.45% C steel using a multi-tasking machine. In the FSB process, the burnishing tool rotates at a high-revolution speed. The thin surface layer is rubbed and stirred as the temperature is increased and decreased. With the FSB process, high hardness or compressive residual stress can be obtained on the surface layer. However, when we applied the FSB process using a 3 mm diameter sphere tip shape tool, the surface roughness increased substantially ($R_a$ = 20 μm). We therefore used four types of tip shape tools to examine the effect of burnishing tool tip radius on surface roughness, hardness, residual stress in the FSB process. Results indicated that the surface roughness was lowest ($R_a$ = 10 μm) when the tip radius tool diameter was large (30 mm).

**Keywords:** friction stir burnishing; surface enhancement; residual stress; surface roughness; strain

## 1. Introduction

Mechanical parts need to have long life and high performance. This is especially true of cylindrical bars and gears that are used in power transmissions. Such parts must have both high abrasion resistance and high fatigue strength. Then the surface enhancement technologies have been used to provide a new surface characteristic without losing the characteristics of the base material. There are various methods of surface enhancement. For example, induction hardening [1,2] and shot peening [3–5] are very popular. High strength and abrasion resistance can be given by the induction hardening; the surface is hardened while the toughness in the core is maintained. Laser hardening process [6,7] is also employed to make high quality hardened layer on the limited area of the workpiece. These methods are used for auto parts and various kinds of mechanical parts. On the other hand, thermal-sprayed coatings [8] or blasting process [9] can make additional layers using different material from the base metal to get higher hardness, wear resistance, thermal resistance, or corrosion resistance.

Metal cutting is one of the general machining methods used to achieve high accuracy and productivity for metal parts. However, the tensile residual stress often induced within the surface layer after machining and the tensile residual stress weakens the fatigue strength [10]. Therefore, shot peening is often applied after cutting in order to induce the compressive residual stress and higher hardness within the surface layer [6,7]. However, since the shot peening process is an additional process after the machining process, some kinds of combined process with the cutting process is requested from industry to improve the productivity.

One solution is a laser hardening process [6,7], as mentioned before. Laser head can be mounted on the machine tools such as a lathe, then the laser hardening can be executed just after the turning on the lathe. The burnishing processes are also one of the solutions. Most common shape of the burnishing tool is ball or roller. Ball/roller burnishing tool can also be mounted on the machine tools, such as the lathe and the drilling machine. The ball/roller burnishing tool presses the work piece surface and move then the workpiece surface plastically deforms without friction. As the result, surface roughness is improved and compressive residual stress is induced, at same time work-hardening occurs on the surface [11]. Slami et al. applied the ultrasonic burnishing process for Co-Cr and stainless steel components made by additive manufacturing [12]. By the ultrasonic burnishing process surface roughness could be improved and processed surface is hardened.

As the other solution, a friction stir processing [13–18], which is basically based on the similar process of friction stir welding, has been proposed as a novel surface-enhancement technology. In this process, frictional heat and large strain produce grain refinement and a hardened layer. It can be combined with machining processes that use a rotating tool, such as machining center or multi-tasking machine. It can be possible to achieve surface enhancement immediately after the cutting process using general cutting machine tools, additional processes could be avoided, improving manufacturing efficiency.

One of the author of this paper proposed friction stir burnishing (FSB) [19–21] as one of the frictional stir process. Figure 1 contains a schematic of the FSB process. In a thin surface layer, large plastic strain and frictional heat are induced locally, because a sphere-shaped burnishing tool rotating at high rotational speed rubs the surface of a workpiece. This makes the surface temperature rise and fall within a short time period. Then, metallic grain refinement and martensitic transformation could be produced on carbon steel.

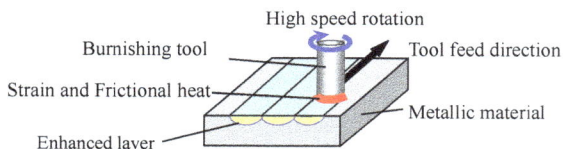

**Figure 1.** Scheme of the friction stir burnishing.

We reported that high hardness (600 HV) and compressive residual stress (−400 MPa) could be obtained by the FSB process on 0.45% C steel [22]. However, when a sphere tip shape tool with a 3 mm radius was applied for the FSB process, surface roughness became very large ($R_a$ = 20 μm) for the machine parts surface. Tanaka et al. proposed combination machining [23], after the FSB process and finish cutting process was conducted again, and showed that a superior surface roughness could be obtained without changing the characteristics of the enhanced layer. However, the additional finish cutting is not efficient. In addition, surface characteristics might change due to the finishing cutting process, depending on the cutting conditions.

Therefore, this study aims to achieve both surface enhancement and superior surface roughness using only the FSB process. In this paper, we used four different burnishing tool tip radii to research the effect of tip shape. Our objective is to clarify the effect of tool tip radius on surface roughness, hardness, and residual stress within the surface layer.

## 2. Materials and Methods

Figure 2 shows the FSB tool assembly. This tool consists of a burnishing pin made of cemented carbide. The burnishing pin was attached at the end of the tool top. A spring was inserted into the tool shank so that the spring preload could be controlled. It was then possible to apply a constant axial force [19]. The experimental setup is shown in Figure 3. The FSB tool was mounted to the tool spindle of a multi-tasking machine (INTEGREX 200-IIIST, Mazak, Oguchi, Japan), and the cylindrical workpiece was set to the work spindle. The workpiece material was 0.45% C steel. The high-speed rotating tool moved in the axial direction of the workpiece rotating at a low speed. The tool path was helical.

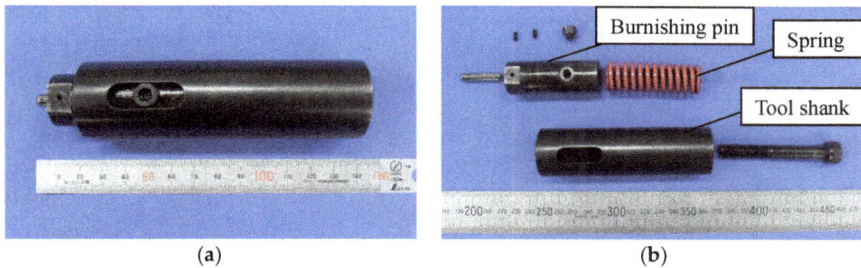

(a)                                            (b)

**Figure 2.** Tool for friction stir burnishing (FSB): (**a**) External view; and (**b**) Components.

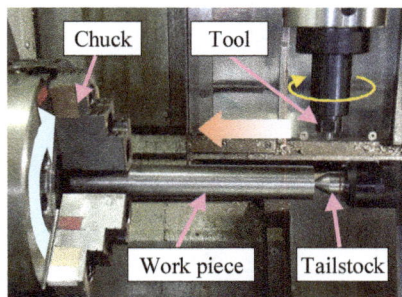

**Figure 3.** The experimental test stand.

After the FSB process was completed, we measured the surface roughness and residual stress on the processed surface. The measurement point was around fourth turn of the helical path. Measurement directions were circumferential and axial on the cylindrical workpiece. The X-ray diffraction method was used for residual stress measurement. Five levels of X-ray incident angles: $0°$, $5°$, $10°$, $15°$, and $20°$, were used for measurement. A part of the machined surface layer was excised and mounted on resin. It was then ground using SiC paper, and polished with a diamond compound. After polishing, samples were etched using nital solution (5% nitric acid in ethanol). Subsurface microstructural analyses were conducted using an optical microscope. Vickers microhardness measurements were conducted with a load of 0.245 N for 20 s. Hardness on the hardened layer was measured at six points on fourth turn of the helical path and averaged.

The FSB process was applied using different tip radius burnishing tools, and surface roughness, hardness, and residual stress were compared. Figure 4 shows the FSB tool tips, and Table 1 lists the specifications of the burnishing tool. The burnishing tool was a 6-mm diameter cylindrical bar, and the tip was machined to sphere shape. In this study, four levels of tool tip radius were used for the FSB process: $R = 3$ mm, 10 mm, 20 mm, and 30 mm. Table 2 presents the processing conditions. The FSB process was applied under the condition of tool thrust force $P = 750$ N, tool spindle speed $S = 10{,}000$ min$^{-1}$, helical pitch $p = 2.5$ mm, and tool feed rate $F = 200$ mm/min. The processing conditions were determined based on our previous study [22] using the tool with $R = 3$ mm. In our previous study, it was shown that tool thrust force $P$ and tool rotation speed $S$ affected strain and heat generation. Lager thrust force ($P$ = 750 N) and higher tool rotation speed ($S$ = 10,000 min$^{-1}$) increased the strain and the heat generation. As a result, phase transition was occurred on processed surface then hardened layer of 600 HV could be obtained. On the other hand, under the condition of smaller thrust force ($P$ = 500 N) and lower tool rotation speed ($S$ = 1000 min$^{-1}$), heat generation was lower. Then, the phase transformation did not occur and the work hardened layer was formed on the surface, but its hardness was 300 HV and was lower than that mentioned above. However, the processed surface was plastically deforms, as a result the compressive residual stress was induced. In addition, higher tool feed rate $F$ decreased the heat generation in per unit tool moving length. Then, the thickness of the hardened layer decreased. When considering the above background, processing conditions in this paper was selected based on the conditions where a hardened layer of 600 HV and 500 μm thickness could be obtained when using the tool with $R$ = 3 mm and the effect of the tool tip radius, which could affect the surface deformation and the heat generation, was mainly studied.

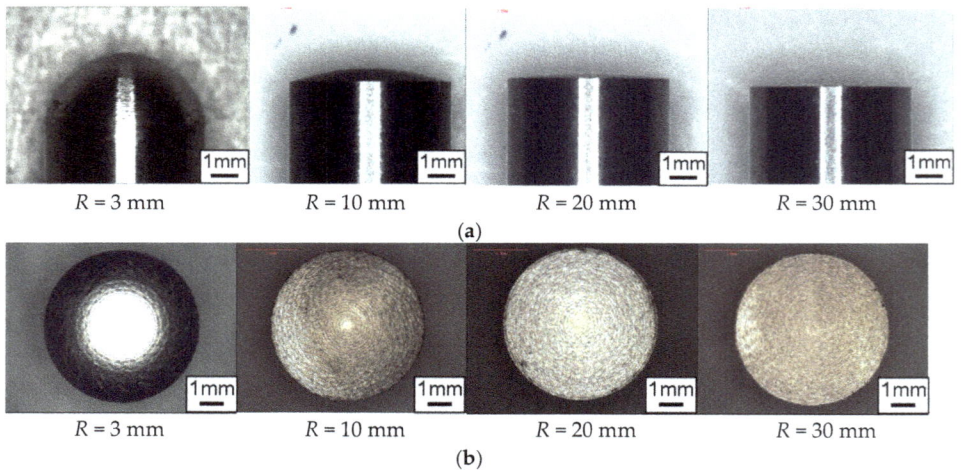

**Figure 4.** Tip shape of tool: (**a**) Side view; (**b**) Over top view.

**Table 1.** Specification of burnishing tool.

| Tool Tip Shape | Tool Radius $R$ (mm) | | | | Tool Material | Tool Dimension (mm) |
|---|---|---|---|---|---|---|
| Sphere | 3 | 10 | 20 | 30 | Cemented carbide | φ6 × 30 |

**Table 2.** Processing condition.

| Tool Thrust Force $P$ (N) | Tool Spindle Speed $S$ (min$^{-1}$) | Helical Pitch $p$ (mm) | Feed Rate $F$ (mm/min) | Tool Material | Workpiece Material |
|---|---|---|---|---|---|
| 750 | 10,000 | 2.5 | 200 | Cemented carbide | 0.45% C steel |

## 3. Results and Discussion

Figure 5a is the photo of $R$ = 3 mm tool taken from the side direction as the typical surface profile of tool after the FSB and Figure 5b shows schematic of tool wear. Wear is evident on the tool tips. It appears that the tool was worn at outer edges of the part where the tool tip is considered to have contacted the work piece, and the center of the worn area is slightly sharp. Figure 6 contains photos of tool tips top view after the FSB process was conducted. The diameter of this area was different depending on tool-tip radius; diameters were larger under the condition of a larger tool tip radius. As shown in Figure 7, the depth of the indentation differed depending on the tool tip radius, and this was thought to be the reason for the difference in the diameter of the contact area between the tool and workpiece. Table 3 lists measurements of contact area diameter and the calculated results for indentation depths. The depth of indentation increased with a smaller tool tip radius. Therefore, it is considered that friction stir affects a deeper layer under the condition of smaller tool tip radius.

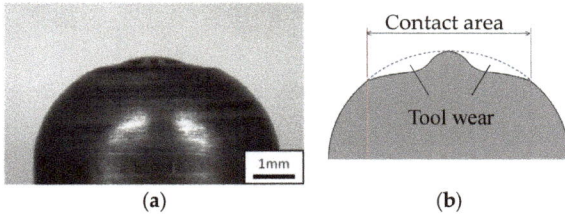

**Figure 5.** Tool tip shape after processing: (**a**) Side view of $R = 3$ mm tool; and (**b**) Schematic of tool wear.

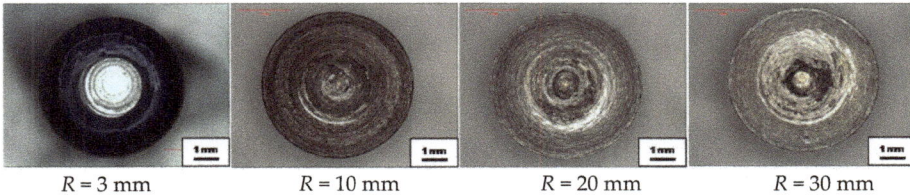

| $R = 3$ mm | $R = 10$ mm | $R = 20$ mm | $R = 30$ mm |

**Figure 6.** Over top view of tool tip after FSB processing.

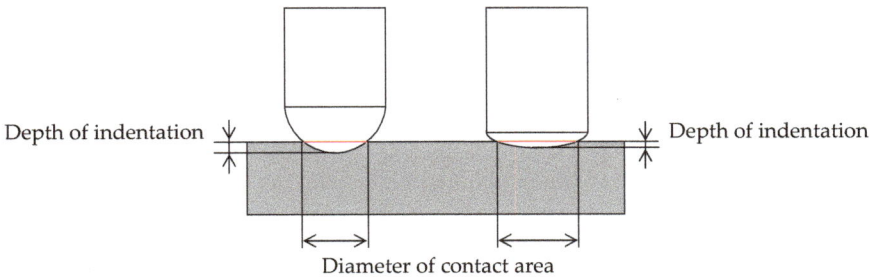

**Figure 7.** Schematics of contact area.

**Table 3.** Diameter of contact area and depth of indentation.

| Tool Radius $R$ (mm) | Diameter of Contact Area (mm) | Depth of Indentation (mm) |
|---|---|---|
| 3 | 2.4 | 0.24 |
| 10 | 2.9 | 0.10 |
| 20 | 3.4 | 0.07 |
| 30 | 3.8 | 0.06 |

Figure 8 shows the volume that stirred by the FSB tool. As shown in Figure 8a, the FSB tool indentation part moves with tool feed, and the shape of the volume that the tool passes during one rotation can be expressed, as shown in Figure 8b. This part is stirred by the burnishing tool. The volume of the stirred part is the same as the semi-cylindrical part shown in Figure 8c. The volume of the stirred part can thus be calculated from the tool tip radius, depth of indentation, and tool moving distance during one rotation. Figure 9 shows the results of the calculated stirred volume during one tool rotation. The stirred volume was larger when a smaller tip radius tool was used.

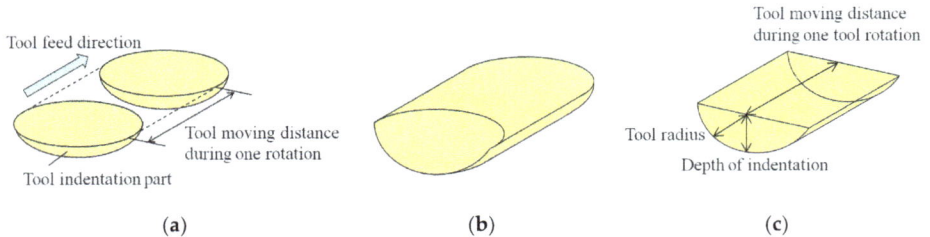

(a)   (b)   (c)

**Figure 8.** Schematic of stirred volume by the FSB tool: (**a**) Tool indentation part and tool movement; (**b**) The volume that is stirred by the tool passing during one rotation; and (**c**) Calculation of stirred volume during one tool rotation.

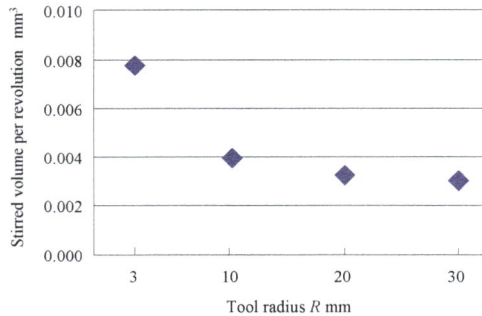

**Figure 9.** Stirred volume per revolution.

Figure 10 contains photos of the processed surface. It can be seen that the color of the surface changed to bluish brown under all conditions. This indicates that an oxide film covered the surface. Regular striations like circular arcs are also visible on the surface. The processed surface is not flat, and a large height difference can be seen between each turn of the helical path, especially when a smaller tip radius tool was used. Figure 11 shows the surface roughness that was measured in the circumferential and axial directions of the workpiece. Surface roughness along the circumferential direction ranged from $R_a$ = 7–10 μm. Surface roughness along the axial direction was larger than in the circumferential direction, and increased with a decreased tool tip radius. The largest surface roughness was $R_a$ = 20 μm when tool tip radius $R$ was 3 mm; the smallest was $R_a$ = 10 μm when the tool tip radius $R$ was 30 mm.

Figure 12 shows etched photomicrographs of a cross section of the processed layer vertical to the tool feed direction. It also shows the surface profile vertical to the tool feed direction. It can be seen that the processed surface profile is not flat; raised and depressed areas were generated on the processed surface when a smaller tip radius tool was used. In the FSB process, the side where the tool peripheral moving direction is the same as the tool feed direction is defined as the advancing side (AS), and the side where the tool peripheral moving direction and the tool advance direction are opposite is defined as the retreating side (RS). The raised area was on the retreating side, and the depressed area on the advancing side.

In the FSB process, it is considered that the material contacting the tool surface is forced to move with the frictional stir action. The relative speed between tool surface and workpiece is higher at the advancing side than the retreating side, because the tool peripheral speed relative to the test piece on the advancing side is the sum of the tool rotation speed and feed speed. It is therefore considered that the material moving from the advancing side piles up on the retreating side, while a smaller amount of material moves from the retreating side to the advancing side. As a result, a raised area is generated on the retreating side and a depressed area on the advancing side.

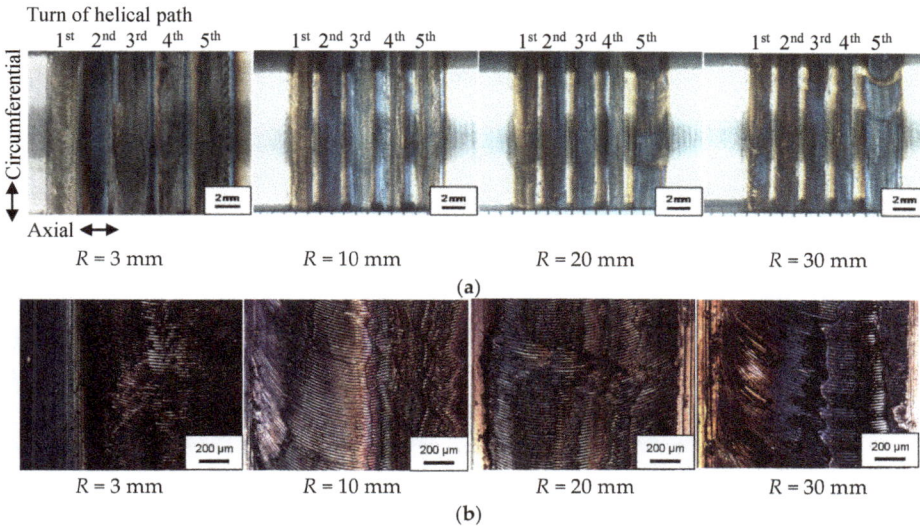

Figure 10. Processed surfaces: (**a**) Overall view; and (**b**) Enlarged view.

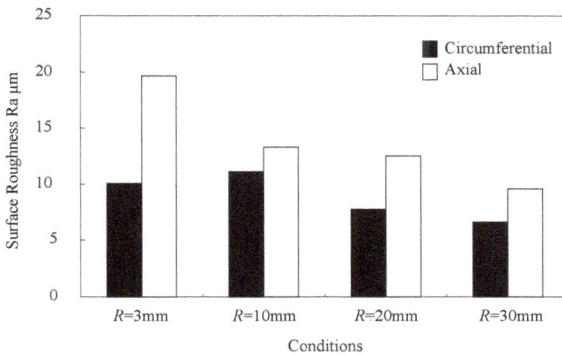

Figure 11. Surface roughness.

Stirred volume becomes large when a smaller tip radius tool is used. The large stirred volume made the height difference on the tool path larger. Then, surface roughness increased with a smaller tip radius tool. In addition, it can be seen in Figure 12 that the processed layer was not etched by the nital. The appearance of the enhanced layer was significantly different from that of the base material, which consisted of ferrite and pearlite.

Figure 13 shows the hardness distribution of the cross section. The hardness value of this layer reached 600 HV. The energy input by the FSB process is consumed as large plastic strain and frictional work, then heat is generated. During the FSB process, the surface temperature increased via the extremely high strain and frictional work, and then decreased rapidly. In these situations, a fine-grain martensite is formed in the processed region [22]. The thickness of the hardened layer varied, depending on the tool tip radius, increasing when a smaller tip radius tool was used. This is because the stirred volume during one tool rotation is increased with the smaller tip radius tool, inducing a larger amount of plastic strain to the processing region, and increasing the amount of the heat generation by plastic strain.

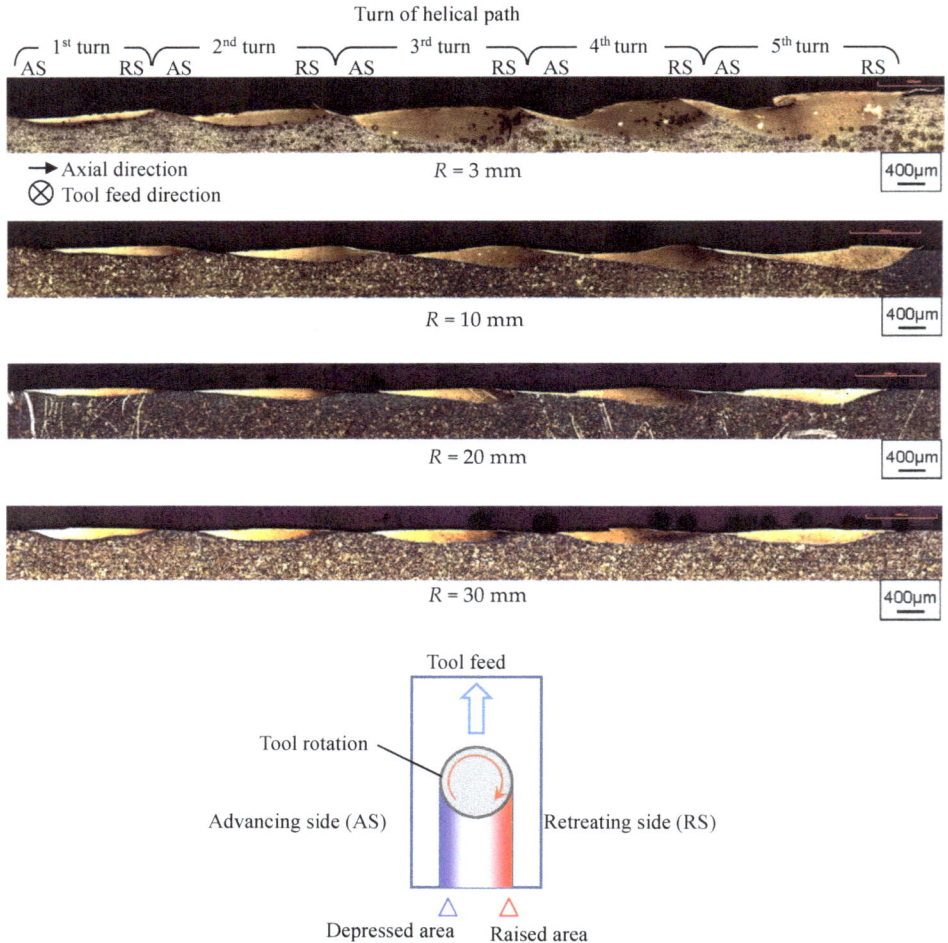

Figure 12. Cross sections of processed surfaces.

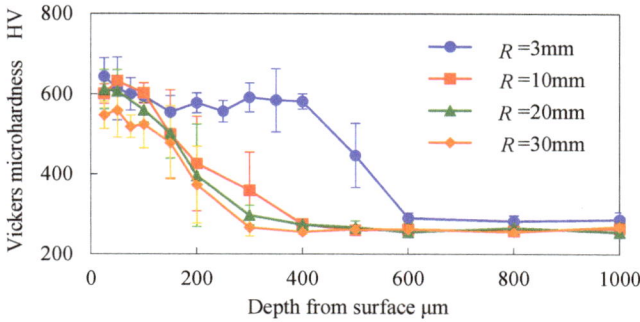

**Figure 13.** Hardness distributions on cross section of processed surfaces.

Figure 14 shows the residual stress measured on the processed surface. The residual stress under the condition $R = 3$ mm and 10 mm was about $-100$ MPa compressive stress; the residual stress turned tensile as the tip radius tool size increased. A smaller tool tip radius is preferable for inducing compressive residual stress. It is considered that larger plastic strain, which stirs and compresses the surface layer with a smaller tip radius tool, leads to the compressive residual stress. On the other hand, the residual stress becomes tensile as the tool tip radius increases. Under all of the processing conditions, it is considered that the surface temperature was high enough for the austenite transformation. The surface temperature then decreased to room temperature. At this point, thermal contraction occurred due to temperature decrease, and tensile residual stress was induced. When a larger tip radius tool was used, the effect of the thermal contraction was larger than the effect of the plastic strain induced by frictional stirring; as a result, the residual stress became tensile.

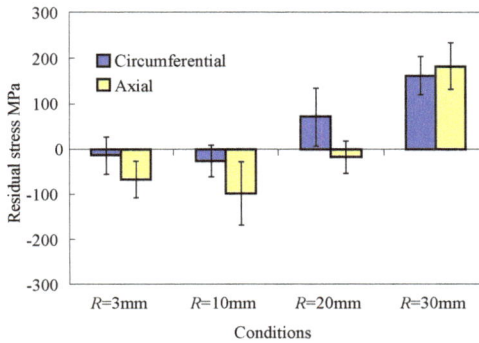

**Figure 14.** Residual stresses on processed surface.

## 4. Conclusions

We studied the FSB process using four levels of tip radius burnishing tools and investigated their effect on surface roughness, hardness and residual stress for 0.45% C steel shaft. The conclusions in this study can be summarized as follows.

- Surface roughness increased under the condition of a smaller tool tip radius. The largest surface roughness was $R_a = 20$ μm under the condition of a tool tip radius of $R = 3$ mm, and the smallest surface roughness was $R_a = 10$ μm under the condition of tool tip radius $R = 30$ mm.
- The thickness of the hardened layer increased as the tool tip radius decreased. The hardness value of the processed layer reached 600 HV.

- The residual stress on the processed surface was compressive when a smaller tip radius tool was used, and the residual stress turned tensile when a larger tip radius tool was used.

**Acknowledgments:** This work was supported by JSPS KAKENHI Grant Number JP19360060.

**Author Contributions:** Yoshimasa Takada and Hiroyuki Sasahara conceived and designed the experiments; Yoshimasa Takada performed the experiments; Yoshimasa Takada and Hiroyuki Sasahara analyzed the data; Yoshimasa Takada wrote the paper.

**Conflicts of Interest:** The authors declare no conflict of interest.

## References

1. Kristoffersen, H.; Vomacka, P. Influence of process parameters for induction hardening on residual stresses. *Mater. Des.* **2001**, *22*, 637–644. [CrossRef]
2. Coupard, D.; Palin-luc, T.; Bristiel, P.; Ji, V.; Dumas, C. Residual stress in surface induction hardening of steel: Comparison between experiment and simulation. *Mater. Sci. Eng. A* **2008**, *487*, 328–339. [CrossRef]
3. Torres, M.A.S.; Voorwald, H.J.C. An evaluation of shot peening, residual stress and stress relaxation on the fatigue life of AISI 4340 steel. *Int. J. Fatigue* **2002**, *24*, 877–886. [CrossRef]
4. Bagherifard, S.; Slawik, S.; Fernández-Pariente, I.; Pauly, C.; Mücklich, F.; Guagliano, M. Nanoscale surface modification of AISI 316L stainless steel by severe shot peening. *Mater. Des.* **2016**, *102*, 68–77. [CrossRef]
5. Tosha, K. History and Future of Shot Peening. *Mater. Jpn.* **2008**, *47*, 134–139. [CrossRef]
6. Ding, H.; Shin, Y.C. Laser-assisted machining of hardened steel parts with surface integrity analysis. *Int. J. Mach. Tool. Manuf.* **2010**, *50*, 106–114. [CrossRef]
7. Klocke, F.; Schulz, M.; Gräfe, S. Optimization of the laser hardening process by adapting the intensity distribution to generate a top-hat temperature distribution using freeform optics. *Coatings* **2017**, *7*, 77. [CrossRef]
8. Zhang, G.; Li, D.; Zhang, N.; Zhang, N.; Duan, S. Thermal-sprayed coatings on bushing and sleeve-pipe surfaces in continuous galvanizing sinking roller production line applications. *Coatings* **2017**, *7*, 113. [CrossRef]
9. O'Sullivan, C.; O'Hare, P.; Byrne, G.; O'Neill, L.; Ryan, K.B.; Crean, A.M. A modified surface on titanium deposited by a blasting process. *Coatings* **2011**, *1*, 53–71. [CrossRef]
10. Sasahara, H. The effect on fatigue life of residual stress and surface hardness resulting from different cutting conditions of 0.45%C steel. *Int. J. Mach. Tool. Manuf.* **2005**, *45*, 131–136. [CrossRef]
11. Mahajan, D.; Tajane, R. A review on ball burnishing process. *Int. J. Sci. Res. Publ.* **2013**, *3*, 1–8.
12. Salmi, M.; Huuki, J.; Ituarte, I.F. The ultrasonic burnishing of cobalt-chrome and stainless steel surface made by additive manufacturing. *Prog. Addit. Manuf.* **2017**, 31–41. [CrossRef]
13. Mishra, R.S.; Ma, Z.Y. Friction stir welding and processing. *Mater. Sci. Eng. R Rep.* **2005**, *50*, 1–78. [CrossRef]
14. Grewal, H.S.; Arora, H.S.; Agrawal, A. Surface modification of hydroturbine steel using friction stir processing. *Appl. Surf. Sci.* **2013**, *268*, 547–555. [CrossRef]
15. Nia, A.A.; Omidvar, H.; Nourbakhsh, S.H. Effects of an overlapping multi-pass friction stir process and rapid cooling on the mechanical properties and microstructure of AZ31 magnesium alloy. *Mater. Des.* **2014**, *58*, 298–304. [CrossRef]
16. Lorenzo-Martin, C.; Ajayi, O.O. Rapid surface hardening and enhanced tribological performance of 4140 steel by friction stir processing. *Wear* **2015**, *332–333*, 962–970. [CrossRef]
17. Saito, N.; Shigematsu, I. Friction stir processing—A new technique for microstructure control of metallic materials. *J. Jpn. Inst. Light Met.* **2007**, *57*, 492–498. [CrossRef]
18. Fujii, H.; Yamaguchi, Y.; Kikuchi, T.; Kiguchi, S.; Nogi, K. Surface hardening of two cast irons by friction stir processing. *J. Phys. Conf. Ser.* **2009**, *165*, 012013. [CrossRef]
19. Sasahara, H.; Kiuchi, S.; Yata, T.; Murase, H.; Tominaga, K. Generation of surface hardened layer on 0.45%C steel by frictional stir burnishing. In Proceedings of the 9th Global Congress on Manufacturing and Management, Gold Coast, Australia, 12–14 November 2008.
20. Kiuchi, S.; Sasahara, H. Temperature history and metallographic structure of 0.45%C steel processed by friction stir burnishing. In Proceedings of the 5th Leading Edge Manufacturing 21st Century, Omiya, Japan, 2–4 December 2009; pp. 359–364.

21.   Kiuchi, S.; Sasahara, H. Temperature History and Metallographic Structure of 0.45%C Steel Processed by Frictional Stir Burnishing. *J. Adv. Mech. Des. Syst.* **2010**, *4*, 838–848. [CrossRef]

22.   Takada, Y.; Sasahara, H. Frictional stir burnishing on double helical path to satisfy both high hardness and compressive residual stress. *Trans. Jpn. Soc. Mech. Eng.* **2015**, *81*, 15-00350. [CrossRef]

23.   Tanaka, R.; Okada, R.; Nakagawa, T.; Furumoto, T.; Hosokawa, A.; Ueda, T. Creation of modified surface layer with superior roughness by combination machining. *Trans. Jpn. Soc. Mech. Eng. Ser. C* **2012**, 3605–3614. [CrossRef]

*coatings*

MDPI

*Article*

# Influence of Thickness of Multilayered Nano-Structured Coatings Ti-TiN-(TiCrAl)N and Zr-ZrN-(ZrCrNbAl)N on Tool Life of Metal Cutting Tools at Various Cutting Speeds

**Alexey Vereschaka [1],\*, Elena Kataeva [2], Nikolay Sitnikov [3], Anatoliy Aksenenko [4], Gaik Oganyan [5] and Catherine Sotova [6]**

1   Department of Mechanical Engineering, Moscow State Technological University STANKIN, Moscow 127055, Russia

2   University Administration, Moscow State Technological University STANKIN, Moscow 127055, Russia; rector@stankin.ru

3   Department of Solid State Physics and Nanosystems, National Research Nuclear University MEPhI, Moscow 115409, Russia; sitnikov_nikolay@mail.ru

4   Material Properties Research Laboratory, Moscow State Technological University STANKIN, Moscow 127055, Russia; a.aksenenko@lism-stankin.ru

5   Material Cutting Technology Laboratory, Moscow State Technological University STANKIN, Moscow 127055, Russia; svartrans88@yandex.ru

6   Department of Composite Materials, Moscow State Technological University STANKIN, Moscow 127055, Russia; e.sotova@stankin.ru

\*   Correspondence: ecotech@rambler.ru; Tel.: +7-916-910-0413

Received: 24 December 2017; Accepted: 16 January 2018; Published: 23 January 2018

**Abstract:** This paper considers the influence of thickness of multilayered nano-structured coatings Ti-TiN-(TiCrAl)N and Zr-ZrN-(ZrCrNbAl)N on tool life of metal cutting tools at various cutting speeds ($v_c$ = 250, 300, 350 and 400 m·min$^{-1}$). The paper investigates the basic mechanical parameters of coatings and the mechanism of coating failure in scratch testing depending on thickness of coating. Cutting tests were conducted in longitudinal turning of steel C45 with tools with the coatings under study of various thicknesses (3, 5, and 7 μm), with an uncoated tool and with a tool with a "reference" coating of TiAlN. The relationship of "cutting speed $v_c$—tool life T" was built and investigated; and the mechanisms were found to determine the selection of the optimum coating thickness at various cutting speeds. Advantages of cutting tools with these coatings are especially obvious at high cutting speeds (in particular, $v_c$ = 400 m·min$^{-1}$). If at lower cutting speeds, the longest tool life is shown by tools with thicker coatings (of about 7 μm), then with an increase in cutting speed (especially at $v_c$ = 400 m·min$^{-1}$) the longest tool life is shown by tools with thinner coating (of about 3 μm).

**Keywords:** multilayered nano-structured coatings; tool life; metal cutting tools; scratch testing

---

## 1. Introduction

Wear-resistant coatings are actively and successfully applied to modify the superficial layer of tool materials and thus to increase performance properties of cutting tools. On the one hand, the use of modifying coatings makes it possible to increase tool life, while on the other hand, that can significantly increase cutting modes and the cutting speed [1–5]. Coating thickness is an important indicator that significantly affects the performance properties of metal cutting tools. The choice to select the optimum coating thickness for different machining conditions was studied by a number of researchers. In particular, Klocke et al. [6] noted that carbide cutting tools with thicker PVD coatings are characterized by longer tool life and that contributes to reduction of production costs. Meanwhile, Messier et al. [7] showed that when a monolayered coating is deposited, its grains grow with increasing coating thickness. Accordingly, the superficial hardness of monolayered coating will decrease with increase in its thickness [6]. It can also be assumed that the mechanical strength of thin coatings will be higher than that of thicker coatings. It is shown that nominal superficial hardness, superficial yield, and maximum superficial strength decrease with an increase in coating thickness [6]. Bouzakis et al. [8–11] studied the influence of thickness for coating (TiAl)N (coating with thickness of 2–10 μm was studied) on the tool life of a carbide tool when turning steel at various cutting modes. It is found that tool life improves with increased coating thickness. Maruda et al. [12] studied finish turning of steel for different cooling conditions: dry cutting, minimum quantity cooling and lubrication. Krolczyk et al. studied tools which were coated with $Al_2O_3$ [13] and $TiN$-$Al_2O_3$-$Ti(C,N)$ [14]. Recently, the properties of multicomponent coatings, sometimes called "highly entropic" coatings, have been extensively studied. In particular, the following coatings were investigated (Ti,V,Cr,Zr,Hf)N [15,16], (Al,Cr,Ta,Ti,Zr)N [17], (Al,Cr,Nb,Si,Ti,V)N [18], and (Zr,Nb,Cr,Al)N [19]. In these studies, it was shown that the addition of alloying elements (in particular, Nb, Cr, Nf, Ta) to the Zr-N, Al-N or Ti-N nitride systems reduces the average cavitation velocity and abrasive wear of the coatings.

Proceeding from the above, it can be noted there is conflicted opinion on coating thickness. On the one side, a number of authors argue that tool life improves with an increase in coating thickness (up to 10 μm), while other authors note a marked decrease in the performance properties of a coating as its thickness increases. Meanwhile, the influence of thickness of a multilayered nano-structured coating on tool life was not in fact studied. The purpose of this study was to investigate the influence of wear-resistant layer thickness and elemental composition of a coating on tool life at various cutting speeds ($v_c$ = 250, 300, 350, and 400 m·min$^{-1}$ were considered).

## 2. Materials and Methods

For the comparative tests, two types of multilayered nano-structured coatings were selected: Ti-TiN-(TiCrAl)N and Zr-ZrN-(ZrCrNbAl)N, each with three different thickness (3, 5, and 7 μm). These coatings were selected as the most effective ones in accordance with the results of previous tests [20–25]. The monolayered non-nano-structured coating TiAlN with a thickness of 4 μm, as well as carbide uncoated insert, were selected for comparison. These coatings were deposited on carbide inserts SNUN ($\gamma = -8°$, $\alpha = 8°$, $K = 45°$, $\lambda = 0$, $R = 0.8$ mm). For coating deposition, the filtered cathodic vacuum-arc deposition (FCVAD) VIT-2 unit (MSTU STANKIN, Moscow, Russia) [20,23] was used. The cutting tests were carried out at the following cutting conditions: $f = 0.2$ mm/rev; $a_p = 1.0$ mm; $v_c = 250, 300, 350,$ and $400$ m·min$^{-1}$. Tool failure criterion was flank wear land $V_B = 0.4$ mm. The tests were conducted for three tips in each mode, and then arithmetic mean of tool life was determined. For microstructural studies of samples, a SEM FEI Quanta 600 FEG (Materials & Structural Analysis Division, Hillsboro, OR, USA) was used. The microhardness (HV) of coatings was measured by using the method of Oliver and Pharr [26], at a fixed load of 300 mN. The adhesion characteristics were studied on a Nanovea scratch tester (Micro Scratch, Nanovea, Irvine, CA, USA). The tests were carried out with the load linearly increasing from 0.05 to 40 N. Two points were registered: $L_{c1}$—the point for the formation of the first cracks in the coating, $L_{c2}$—the point of complete destruction of the coating.

## 3. Results and Discussion

The results of the main parameters of coatings are shown in Table 1.

In the course of the investigation of adhesion bond strength by scratch testing, not only quantitative indicators, but also the very fracture pattern is important. Let us consider in more detail the differences in coating fracture patterns depending on their thicknesses and structures. Let us consider the differences in the fracture patterns for multilayered nano-structured coatings Ti-TiN-(TiCrAl)N of various thicknesses and monolayered TiAlN coating. When TiAlN coating fails, extensive areas of spallation from substrate are formed at the final stage (Figure 1a). Despite the fact that this coating is monolayered and that it has no layered nano-structure, longitudinal fractures are also observed during the process of its destruction (Figure 1b). Cross cracks can also be seen (Figure 1b). In this case, there is classic brittle fracture, and that corresponds to high hardness and, respectively, low ductility of the coating. The fracture mechanisms for this coating can be classified as recovery spallation and compressive spallation.

**Table 1.** Main parameters of the coatings under study.

| Type of Coating | Coating Thickness (µm) | Microhardness (GPa) | Strength of Adhesion Bond to Substrate | |
|---|---|---|---|---|
| | | | $L_{c1}$ (N) | $L_{c2}$ (N) |
| TiAlN | 4 (±0.8) | 30.3 | 23.2 | 30.1 |
| Ti-TiN-(TiCrAl)N | 3 (±0.7) | 32.7 | 32.6 N | 34.2 |
| Ti-TiN-(TiCrAl)N | 5 (±0.6) | 32.2 | 31.2 | 35.0 |
| Ti-TiN-(TiCrAl)N | 7 (±0.6) | 33.5 | – | 34.8 |
| Zr-ZrN-(ZrCrNbAl)N | 3 (±0.6) | 29.4 | 30.2 | 32.6 |
| Zr-ZrN-(ZrCrNbAl)N | 5 (±0.5) | 30.1 | 31.7 | 33.1 |
| Zr-ZrN-(ZrCrNbAl)N | 7 (±0.5) | 30.2 | – | 33.2 |

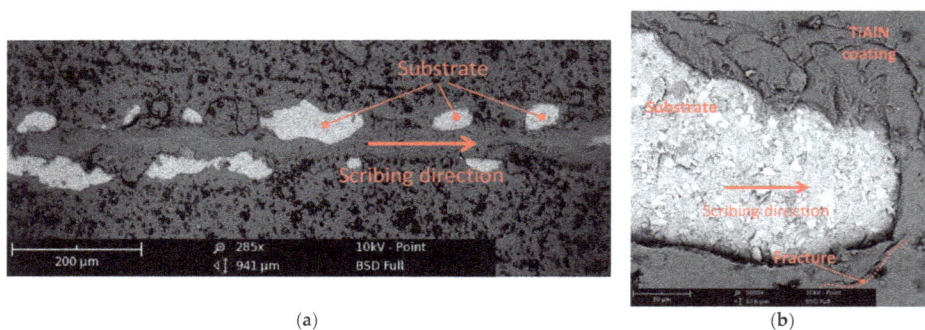

(a)                                              (b)

**Figure 1.** Fracture pattern for monolayered coating TiAlN in scratch testing: (**a**) General view of scribing groove in area of $L_{c1}$–$L_{c2}$; (**b**) Area of coating spallation from substrate.

Meanwhile, a different fracture mechanism is typical for multilayered nano-structured Ti-TiN-(TiCrAl)N coating (thickness of about 3 µm). While this coating is characterized by a hardness even slightly higher than the hardness of the TiAlN coating tested earlier, due to its layered nano-structure, this coating is more ductile and its resistance to brittle fracture is higher. While for TiAlN coating, points of spallation from substrate are mainly formed at edge sections of the groove (Figure 1a), for Ti-TiN-(TiCrAl)N coating (thickness of about 3 µm), such spallation can be mainly observed in a center of a scribing groove. The very fracture pattern for Ti-TiN-(TiCrAl)N coating (thickness of about 3 µm) is mainly determined by ductile mechanisms (Figure 2a), and that is especially noticeable for its outer wear-resistant layer. At the same time, intermediate layer TiN is characterized by more brittle fracture mechanisms (Figure 2b). The fracture mechanism typical for this coating can be classified as gross spallation.

Another fracture mechanism is typical for coating Ti-TiN-(TiCrAl)N (thickness of about 5 µm): there is wedge-shaped spalling and interlayer delamination (Figure 3a). The combination of both ductile and brittle forms of fracture is typical. Spallation occurs not only along the borders of the coating layers, but also along the borders of its nano-layers. Meanwhile, under delamination, there is a slight decrease in internal stresses, and that results in an increase in the threshold of the coating destruction. At higher magnification, it is possible to see the character of crack formation in the coating structure, as well as the role of microdroplets embedded in the coating structure as stress concentrators (Figure 3b).

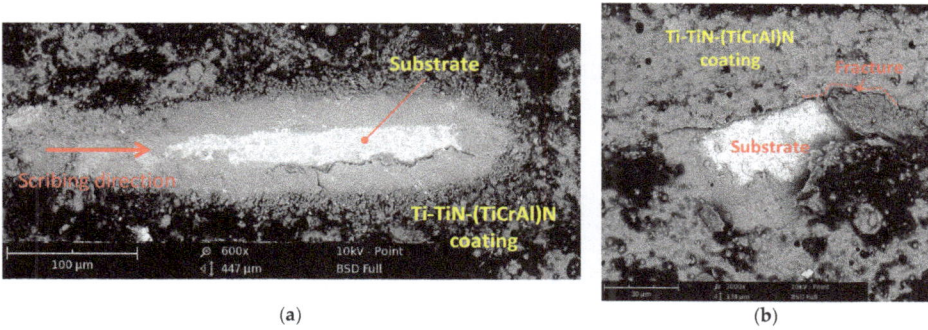

**Figure 2.** Fracture pattern for multilayered nano-structured coating Ti-TiN-(TiCrAl)N (thickness of about 3 μm) in scratch testing: (**a**) General view of scribing groove in area of $L_{c1}$–$L_{c2}$; (**b**) Area of coating spallation from substrate.

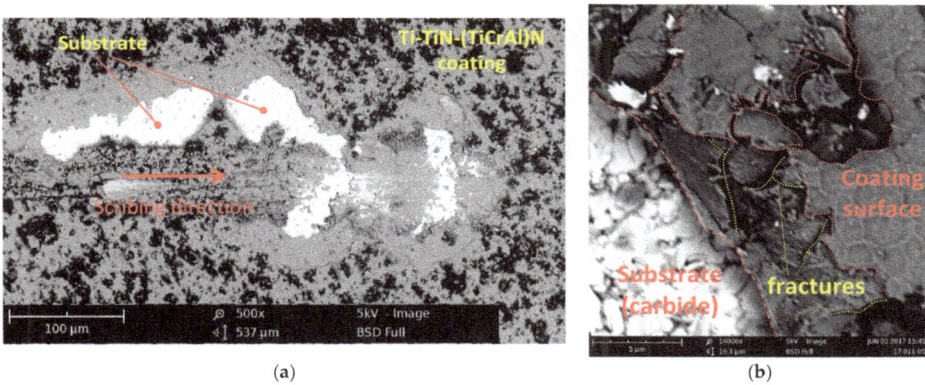

**Figure 3.** Fracture pattern for multilayered nano-structured coating Ti-TiN-(TiCrAl)N (thickness of about 5 μm) in scratch testing: (**a**) General view of scribing groove in area of $L_{c1}$–$L_{c2}$; (**b**) Area of coating fracture.

Finally, the fracture pattern for Ti-TiN-(TiCrAl)N (thickness of about 5 μm), the thickest coating of those under study, is characterized by pronounced brittle fracture with active cracking. It is important to note that for this coating, no $L_{c1}$ typical for appearance of the first cracks is registered, the destruction occurs immediately and completely. Herewith, it can be seen that intermediate layer TiN retains adhesion to the substrate, and only wear-resistant layer (TiCrAl)N is destroyed (Figure 4a). With higher magnification, a typical 'pattern' can be observed associated with delaminations between nano-sublayers of wear-resistant layer of coating (Figure 4b). These delaminations can inhibit the formation of cross cracks, and that can slightly improve the performance properties of the coating [16].

The structures of coatings on cross-section are shown in Figures 5 and 6.

It can be seen that monolayered TiAlN coating has no nano-structure, while Zr-ZrN-(ZrCrNbAl)N and Ti-TiN-(TiCrAl)N coatings show a clear nano-structure of a wear-resistant layer and a transition layer without nano-structure. An adhesive layer cannot be determined in the figure due to its small thickness (about 20 nm).

(a)                                    (b)

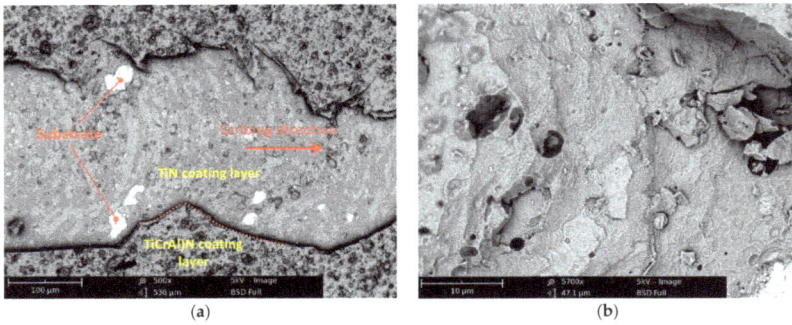

**Figure 4.** Fracture pattern for multilayered nano-structured coating Ti-TiN-(TiCrAl)N (thickness of about 7 μm) in scratch testing: (**a**) General view of scribing groove in area of $L_{c1}$–$L_{c2}$; (**b**) Area of coating fracture.

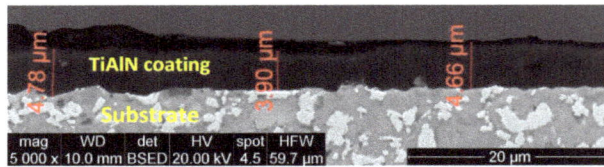

**Figure 5.** Structure on cross-section of coating TiAlN, with thickness 4 (±0.8) μm.

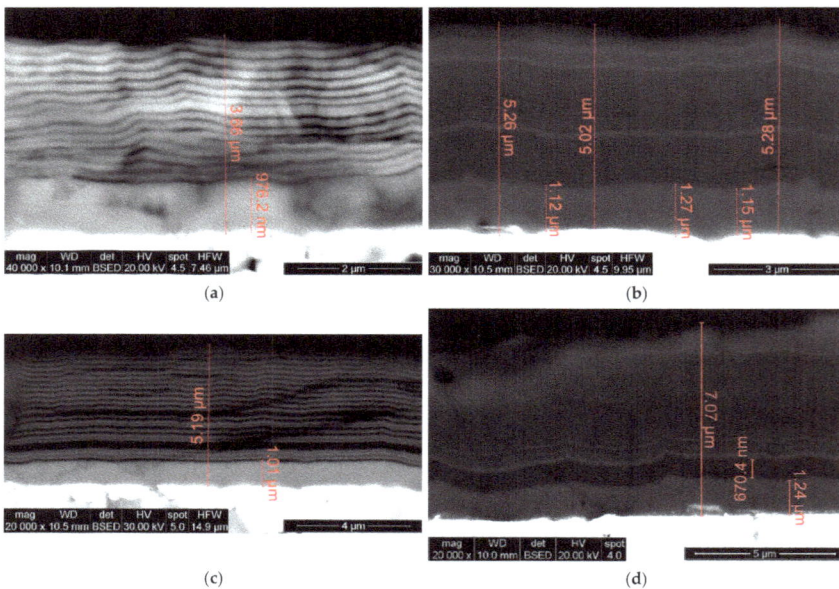

(a)                                    (b)

(c)                                    (d)

**Figure 6.** Structure on cross-section of Zr-ZrN-(ZrCrNbAl)N and Ti-TiN-(TiCrAl)N coatings with a thickness of about 5 μm: (**a**) Zr-ZrN-(ZrCrNbAl)N, thickness 3 (±0.7) μm; (**b**) Ti-TiN-(TiCrAl)N, thickness 5 (±0.5) μm; (**c**) Zr-ZrN-(ZrCrNbAl)N, thickness 5 (±0.6) μm; (**d**) Ti-TiN-(TiCrAl)N, thickness 7 (±0.5) μm.

The results of cutting tests for coated and uncoated and tools under study are shown in Figure 7. On the basis of the results obtained, it can be noted that:

- At a cutting speed of $v_c = 400$ m·min$^{-1}$, an uncoated insert shows excess flank wear after the very first minute of cutting which indicates that uncoated tools cannot be used under these cutting modes. If at cutting speed of $v_c = 250$ m·min$^{-1}$, a tool with monolayered TiAlN coating shows tool life close to durability of multilayered coatings with thickness of about 3 μm, then with increasing cutting speed, tool life of a tool with such coating decreases significantly faster than the tool life of a tool with multilayered nano-structured coating. At cutting speed of $v_c = 400$ m· min$^{-1}$, a tool with monolayered coating TiAlN operates significantly worse than tools with multilayered nano-structured coatings Zr-ZrN-(ZrCrNbAl)N and Ti-TiN-(TiCrAl)N under study.

- If at cutting speed of $v_c = 250$ m·min$^{-1}$ the longest tool life is shown by tools with thicker coatings, then as the cutting speed increases, the picture begins to change and tools with thinner coatings show better results (especially for coating Zr-ZrN-(ZrCrNbAl)N at $v_c = 400$ m·min$^{-1}$). This phenomenon can be explained by the growth of internal stresses in the structure of coating with increasing cutting speed, this process is especially notable in thicker coatings. While there are currently no methods for direct measurement of internal stresses in the structure of coatings several μm thick, there are indirect methods to detect growth of those stresses, at least on a qualitative level. Find more details on the issue in [19].

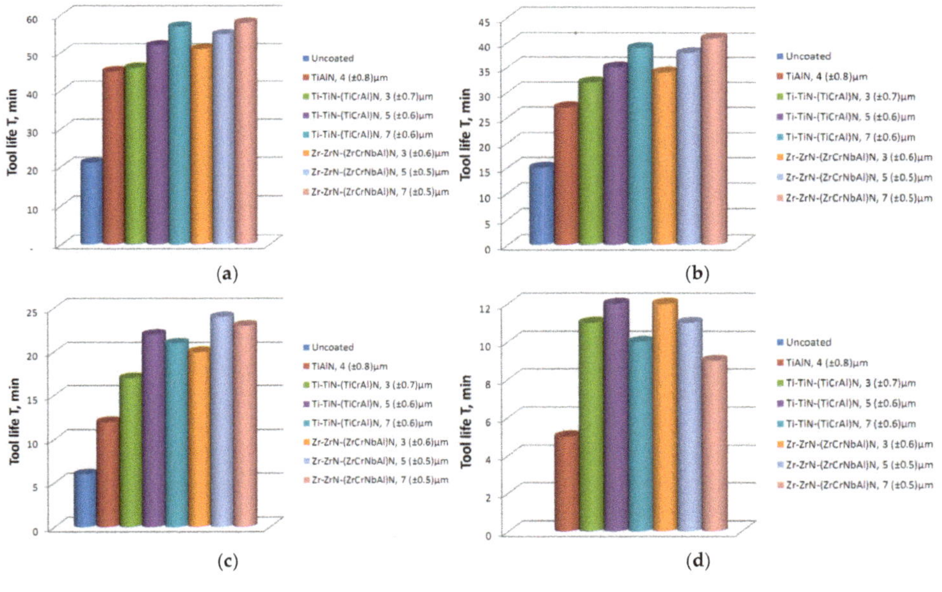

**Figure 7.** Tool life of tools with coatings under study and of uncoated tools at cutting speeds of (**a**) $v_c = 250$, (**b**) $v_c = 300$, (**c**) $v_c = 350$, and (**d**) $v_c = 400$ m·min$^{-1}$ ($f = 0.2$ mm/rev; $a_p = 1.0$ mm, longitudinal turning of steel C45).

Let us now consider the curve of "cutting speed $v_c$—tool life T" (Figure 8). It can be seen that when cutting speed increases, then tool life of a tool with monolayered nano-structured coating TiAlN (Line 1) is most strongly reduced. Cutting tools with multilayered nano-structured coating Ti-TiN-(TiCrAl)N (thickness 5 (±0.6) μm) (Line 3) and Zr-ZrN-(ZrCrNbAl)N (thickness 3 (±0.6) μm) (Line 5) show the lowest reduction in tool life with an increase in cutting speed. Tools with "thicker" coatings Ti-TiN-(TiCrAl)N (thickness 7 (±0.6) μm) (Line 4) and Zr-ZrN-(ZrCrNbAl)N (thickness 7 (±0.5) μm) (Line 7) showed the most noticeable reduction in tool life with an increase in cutting speed. This is especially noticeable for tools with Zr-ZrN-(ZrCrNbAl)N coating (thickness 7 (±0.5) μm) (Line 7), which shows the longest tool life at cutting speed $v_c$ = 250 m·min$^{-1}$, but at a cutting speed of $v_c$ = 400 m· min$^{-1}$ shows the worst result of all the tools with multilayered nano-structured coatings.

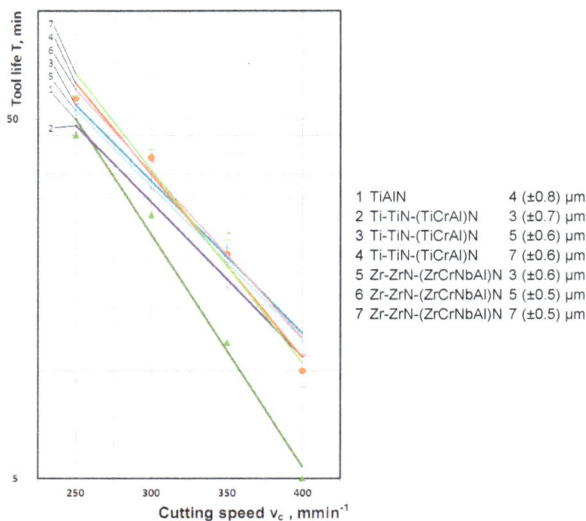

**Figure 8.** Cutting speed $v_c$—tool life T curve at cutting speeds of $v_c$ = 250, 300, 350, and 400 m·min$^{-1}$ ($f$ = 0.2 mm/rev; $a_p$ = 1.0 mm, longitudinal turning of steel C45).

## 4. Conclusions

The use of multilayered nano-structured coatings (in particular, Zr-ZrN-(ZrCrNbAl)N and Ti-TiN-(TiCrAl)N coatings) makes it possible to increase the cutting speed in turning of structural steels. Advantages of cutting tools with these coatings are especially obvious at high cutting speeds (in particular, $v_c$ = 400 m·min$^{-1}$). The longest tool life is shown by tools with thicker coatings (of about 7 μm) at lower cutting speeds, then with an increase in cutting speed (especially at $v_c$ = 400 m·min$^{-1}$) the longest tool life is shown by tools with thinner coatings (of about 3 μm). This phenomenon may be explained by more significant growth of internal stresses in thick coatings with an increase in cutting speeds. High internal stresses result in the formation of internal cracks and interlayer delamination that ultimately leads to the destruction of the coating. Tools with complex multicomponent composition Zr-ZrN-(ZrCrNbAl)N coating show the best results at a speed of $v_c$ = 250 m·min$^{-1}$, however with increasing cutting speed this coating works worse, at a cutting speed of $v_c$ = 400 m·min$^{-1}$ the best result was shown by tools with Ti-TiN-(TiCrAl)N coating. A further development of this research can be the search for patterns that allow a reliable justification of the choice of the composition of coatings and their micro- and nanostructures depending on the material being machined and the cutting regimes (in particular, the cutting speed).

**Acknowledgments:** This research was financed by the Ministry of Education and Science of the Russian Federation (Leading researchers, project 16.9575.2017/6.7).

**Author Contributions:** Alexey Vereschaka and Elena Kataeva conceived and designed the experiments; Nikolay Sitnikov, Anatoliy Aksenenko and Gaik Oganyan performed the experiments; Catherine Sotova analyzed the data; Alexey Vereschaka wrote the paper.

**Conflicts of Interest:** The authors declare no conflict of interest.

# References

1. Vereschaka, A.S. *Working Capacity of the Cutting Tool with Wear Resistant Coatings*; Mashinostroenie: Moscow, Russia, 1993.
2. Hovsepian, P.E.; Ehiasarian, A.P.; Deeming, A.; Schimpf, C. Novel TiAlCN/VCN nanoscale multilayer PVD coatings deposited by the combined high-power impulse magnetron sputtering/unbalanced magnetron sputtering (HIPIMS/UBM) technology. *Vacuum* **2008**, *82*, 1312–1317. [CrossRef]
3. Faga, M.G.; Gautier, G.; Calzavarini, R.; Perucca, M.; Boot, E.A.; Cartasegna, F.; Settineri, L. AlSiTiN nanocomposite coatings developed via Arc Cathodic PVD: Evaluation of wear resistance via tribological analysis and high speed machining operations. *Wear* **2007**, *263*, 1306–1314. [CrossRef]
4. Tkadletz, M.; Schalk, N.; Daniel, R.; Keckes, J.; Czettl, C.; Mitterer, C. Advanced characterization methods for wear resistant hard coatings: A review on recent progress. *Surf. Coat. Technol.* **2016**, *285*, 31–46. [CrossRef]
5. Bouzakis, K.D.; Michailidis, N.; Skordaris, G.; Bouzakis, E.; Biermann, D.; M'Saoubi, R. Cutting with coated tools: Coating technologies, characterization methods and performance optimization. *CIRP Ann. Manuf. Technol.* **2012**, *61*, 703–723. [CrossRef]
6. Klocke, F.; Krieg, T. Coated tools for metal cutting—Features and applications. *Ann. CIRP* **1999**, *48*, 515–525. [CrossRef]
7. Messier, R.; Yehoda, J.E. Geometry of thin-film morphology. *J. Appl. Phys.* **1985**, *58*, 3739. [CrossRef]
8. Bouzakis, K.-D.; Hadjiyiannis, S.; Skordaris, G.; Anastopoulos, J.; Mirisidis, I.; Michailidis, N.; Efstathiou, K.; Knotek, O.; Erkens, G.; Cremer, R.; et al. The influence of the coating thickness on its strength properties and on the milling performance of PVD coated inserts. *Surf. Coat. Technol.* **2003**, *174*, 393–401. [CrossRef]
9. Bouzakis, K.-D.; Hadjiyiannis, S.; Skordaris, G.; Mirisidis, I.; Michailidis, N.; Koptsis, D.; Erkens, G. Milling performance of coated inserts with variable coating thickness on their rake and flank. *CIRP Ann. Manuf. Technol.* **2004**, *53*, 81–84. [CrossRef]
10. Bouzakis, K.-D.; Makrimallakis, S.; Katirtzoglou, G.; Skordaris, G.; Gerardis, S.; Bouzakis, E.; Leyendecker, T.; Bolz, S.; Koelker, W. Adaption of graded Cr/CrN-interlayer thickness to cemented carbide substrates' roughness for improving the adhesion of HPPMS PVD films and the cutting performance. *Surf. Coat. Technol.* **2010**, *205*, 1564–1570. [CrossRef]
11. Skordaris, G.; Bouzakis, K.-D.; Kotsanis, T.; Charalampous, P.; Bouzakis, E.; Lemmer, O.; Bolz, S. Film thickness effect on mechanical properties and milling performance of nano-structured multilayer PVD coated tools. *Surf. Coat. Technol.* **2016**, *307*, 452–460. [CrossRef]
12. Maruda, R.W.; Krolczyk, G.M.; Feldshtein, E.; Nieslony, P.; Tyliszczak, B.; Pusavec, F. Tool wear characterizations in finish turning of AISI 1045 carbon steel for MQCL conditions. *Wear* **2017**, *372*, 54–67. [CrossRef]
13. Krolczyk, G.M.; Nieslony, P.; Legutko, S. Determination of tool life and research wear during duplex stainless steel turning. *Arch. Civ. Mech. Eng.* **2015**, *15*, 347–354. [CrossRef]
14. Królczyk, G.; Gajek, M.; Legutko, S. Predicting the tool life in the dry machining of duplex stainless steel. *Eksploat. Niezawodn. Maint. Reliab.* **2013**, *15*, 62–65.
15. Liang, S.C.; Chang, Z.C.; Tsai, D.C.; Lin, Y.C.; Sung, H.S.; Deng, M.J.; Shieu, F.S. Effects of substrate temperature on the structure and mechanical properties of (TiVCrZrHf)N coatings. *Appl. Surf. Sci.* **2011**, *257*, 7709–7713. [CrossRef]
16. Liang, S.C.; Tsai, D.C.; Chang, Z.C.; Sung, H.S.; Lin, Y.C.; Yeh, Y.J.; Deng, M.J.; Shieu, F.S. Structural and mechanical properties of multi-element (TiVCrZrHf)N coatings by reactive magnetron sputtering. *Appl. Surf. Sci.* **2011**, *258*, 399–403. [CrossRef]
17. Tsai, D.C.; Liang, S.C.; Chang, Z.C.; Lin, T.N.; Shiao, M.H.; Shieu, F.S. Effects of substrate bias on structure and mechanical properties of (TiVCrZrHf)N coatings. *Surf. Coat. Technol.* **2012**, *207*, 293–299. [CrossRef]
18. Chang, S.Y.; Lin, S.Y.; Huang, Y.C.; Wu, C.L. Mechanical properties, deformation behaviors and interface adhesion of (AlCrTaTiZr)Nx multi-component coatings. *Surf. Coat. Technol.* **2010**, *204*, 3307–3314. [CrossRef]
19. Vereschaka, A.A.; Vereschaka, A.S.; Bublikov, J.I.; Aksenenko, A.Y.; Sitnikov, N.N. Study of properties of nanostructured multilayer composite coatings of Ti-TiN-(TiCrAl)N and Zr-ZrN-(ZrNbCrAl)N. *J. Nano Res.* **2016**, *40*, 90–98. [CrossRef]

20. Vereschaka, A.A.; Volosova, M.A.; Batako, A.D.; Vereshchaka, A.S.; Mokritskii, B.Y. Development of wear-resistant coatings compounds for high-speed steel tool using a combined cathodic vacuum arc deposition. *Int. J. Adv. Manuf. Technol.* **2016**, *84*, 1471–1482. [CrossRef]
21. Vereschaka, A.A.; Vereschaka, A.S.; Batako, A.D.; Hojaev, O.K.; Mokritskii, B.Y. Development and research of nanostructured multilayer composite coatings for tungsten-free carbides with extended area of technological applications. *Int. J. Adv. Manuf. Technol.* **2016**, *87*, 3449–3457. [CrossRef]
22. Vereschaka, A.A.; Grigoriev, S.N.; Sitnikov, N.N.; Oganyan, G.V.; Batako, A. Working efficiency of cutting tools with multilayer nano-structured Ti-TiCN-(Ti,Al)CN and Ti-TiCN-(Ti,Al,Cr)CN coatings: Analysis of cutting properties, wear mechanism and diffusion processes. *Surf. Coat. Technol.* **2017**, *332*, 198–213. [CrossRef]
23. Volkhonskii, A.O.; Vereshchaka, A.A.; Blinkov, I.V.; Vereshchaka, A.S.; Batako, A.D. Filtered cathodic vacuum Arc deposition of nano-layered composite coatings for machining hard-to-cut materials. *Int. J. Adv. Manuf. Technol.* **2016**, *84*, 1647–1660. [CrossRef]
24. Vereschaka, A.A.; Grigoriev, S.N.; Sitnikov, N.N.; Batako, A. Delamination and longitudinal cracking in multi-layered composite nano-structured coatings and their influence on cutting tool life. *Wear* **2017**, *390*, 209–219. [CrossRef]
25. Vereschaka, A.A.; Grigoriev, S.N. Study of cracking mechanisms in multi-layered composite nano-structured coatings. *Wear* **2017**, *378*, 43–57. [CrossRef]
26. Oliver, W.C.; Pharr, G.M. An improved technique for determining hardness and elastic modulus using load and displacement sensing indentation. *J. Mater. Res.* **1992**, *7*, 1564–1583. [CrossRef]

**MDPI**

*Article*

# Experimental Evaluation and Modeling of the Damping Properties of Multi-Layer Coated Composites

**Stefano Amadori \*, Giuseppe Cataniaand Angelo Casagrande**

Department of Industrial Engineering—Ciri MaM, University of Bologna, Viale Risorgimento 2,
40136 Bologna, Italy; giuseppe.catania@unibo.it (G.C.); angelo.casagrande@unibo.it (A.C.)
\* Correspondence: stefano.amadori4@unibo.it; Tel.: +39-051-209-3451

Received: 29 December 2017; Accepted: 26 January 2018; Published: 31 January 2018

**Abstract:** In this work, the dissipative properties of different coating solutions are compared and a beam mechanical model, taking into account of dissipative actions at the interface between different layers is proposed. The aim is to find optimal coatings to be employed in the production of composites with high damping properties. The investigated coating layers are obtained from different materials and production processes, and are applied on different metallic substrates. The composite specimens, in the form of slender beams, are tested by means of forced excitation dynamic measurements. Force and displacement experimental data, in a wide range of excitation frequencies, are used to estimate the system damping properties. Homogeneous, uncoated specimens are also tested for comparison. A specific identification procedure is used to identify the specimens stress-strain relationship in the frequency domain. The ratio of the imaginary part and the modulus of the specimen estimated complex frequency response function is considered as a measurement of the damping behaviour. A modified third order multi layered beam model, based on the zig-zag beam theory, is proposed. The model takes into account the contribution to the damping behaviour of the frictional actions and slipping at the interface between layers. Frictional actions are modelled by means of a complex, elasto-hysteretic contribution.

**Keywords:** hysteretic damping; coatings; dynamical measurement; multi-layer beam model; FGM

---

## 1. Introduction

Coating layer technologies are generally used to improve surface hardness, wear strength, thermal resistance, contact friction, with applications in the cutting tool and gas turbine industry [1–6]. Since mainly impulsive, high surface contact forces are expected to be applied to cutting tool components, coating toughness is a major requirement. Plasma Vapour Deposition (PVD) techniques and Chemical Vapour Deposition (CVD) techniques are mainly used in this context [3], generating high residual stresses at the interface between the coating layer and the substrate. In operating conditions, the combination of residual and working load generated stresses can produce coating peeling and surface cracks [2] that compromise the effectiveness of the coating treatment. Coating toughness and maximization of the coating adhesion properties are also a major requirement in other mechanical contexts such as Micro Electro-Mechanical Systems (MEMS) devices, where mainly large flexural displacements and strains, and by consequence high surface stresses, are expected [7]. Residual stress evaluation at the experimental and at the design stage is generally required in this context, and experimental nano-indentation techniques [8,9] and model-based techniques [7] are known.

Coatings can also be employed to increase the global dissipative properties of an industrial component with limited influence on the other mechanical properties [10–12]. Mechanical components with high stiffness, resistance and vibration damping specifications are in great demand for most aerospace, automotive and automation industrial mechanical applications, and some composite solutions are suitable to design components with these properties. In modern high-speed applications, unwanted vibrations may result due to high inertial forces. A vibrational response associated to a small displacement and deformation field but to a wide frequency range may cause an excessive noise level, a decrease in the system efficiency and a shortening of the system service life. By increasing the component dissipative properties, high vibration levels of the contact-free, external surface of thin walled mechanical components such as a mechanical pump, motor or gearing casing may be effectively damped.

Single or multiple layers of a coating material can be deposited or grown in order to produce a finished composite component with specifically designed characteristics including vibrational damping capability. Many different techniques are known [13,14] and in this work specimens obtained by means of the screen printing technology, mainly residual stress free, are compared.

The coating material structure [15–17], the interface structure [18], the temperature dependence [19] are all factors that must be taken into account when studying the influence of coatings on the coated component damping behaviour. The coated component dissipative properties can be significantly tailored by means of the application of coating layers showing high internal hysteresis or with high frictional actions at the interface between the different layers [15,20,21]. Experimental research done by these authors [22,23] and other researchers [13,24] outlined that the application of some coating surface solutions on thin-walled components can increase the vibration damping behaviour, and that this result is mainly due to dissipative actions originating at the interface between the substrate and the coating layer. It is known from literature that dynamic mechanical measurements are an effective experimental tool to study the damping behaviour of coated components and that both forced and free vibrations tests were employed to estimate the dissipative properties of a wide range of coating materials and component geometries by means of comparing the coated and uncoated component dynamical response [25–27]. In this work, the damping behaviour of different coating solutions applied on two different metal substrates, i.e., harmonic steel and Al alloy, are compared. Coated and uncoated specimen dynamic mechanical measurement test results are processed by means of a robust parameter identification and model condensation technique to investigate the effectiveness of the different solutions.

A multi layered beam model, taking into account the frictional actions localized at the interface between the layers, is proposed to help virtually test and find optimal coating solutions, to be applied to a mechanical system casing and to thin-walled components to be used in the high speed automation industry, in order to damp vibration and noise generated in working conditions. Dissipative actions are modeled by relaxing the kinematical displacement continuity at the layer interface and by introducing complex elasto-hysteretic dynamical interface coupling. The effects on the dissipative properties of the distributed constraints modeling boundary conditions are also taken into account. The model is based on zig-zag multi-layer beam theories [28–30], and on layer wise beam theories [31]. High order layer wise beam theories are obtained by modifying the classic Bernoulli-Euler and Timoshenko beam theories in order to deal with composite beams with numerous layers in which the mechanical and geometrical characteristics significantly vary from layer to layer. The advantage of zig-zag theories with respect to other layer wise theories is that the number of state space variables required by the model is low and does not depend on the number of layers. Since large residual stress free, amorphous based structure coatings, mainly deposited by screen printing technologies, are considered in this work, no account is given here with respect to experimental measuring and modeling of residual stresses generated at the interface between two different layers.

## 2. Damping-Oriented Coating Solutions

In this work three innovative, different coating solutions, applied to a slender beam, uniform rectangular cross section test specimen, are compared. Both homogeneous and composite specimens are taken into account. Two different types of metallic substrates are considered, harmonic steel (C67) and Al alloy (Al1000).

Three ceramic coating solutions are proposed, i.e., an alkali activated geopolymer (GP), an alkali activated alumina powder mixture (APM), an alkali activated zirconia powder mixture (ZPM). Ceramic materials, in comparison to metals and polymers, may present superior mechanical, chemical and thermal resistance properties [32].

## 2.1. GP Solution

The geopolymer solution is made by mixing of a commercial metakaolin powder (base) with an aqueous basis binder (activator) prepared from a potassium-hydroxide solution in $H_2O$ with pyrogenic silica solution. The chemical composition of the resulting solution is reported in Table 1.

Potassium was preferred to sodium in the alkaline activator since a better degree of polycondensation can be achieved and because of its ability to provide geopolymeric structures, associated with high mechanical strength [33]. Geopolymers are inorganic polymers formed by linear chains or tridimensional arrays of $SiO_4$ and $AlO_4$ tetrahedra [34]. The geopolymer is produced by mechanical mixing (planetary centrifugal mixer "Thiky Mixer" ARE 500 by THINKY, Tokyo, Japan) of a reactive powder base (Metakaolin Argical M 1200S, IMERYS, Cornwall, UK) with an aqueous basis activator ($H_2O$ solution with potassium-hydroxide of 85% purity and an addition of a 99.8% purity pyrogenic silica solution). The prepared geopolymer has a 2.83 Si/Al ratio composition [34] and is applied to the upper and lower surfaces of the substrate beam by means of screen printing.

**Table 1.** GP composition.

| Base (Metakaolin) | | Activator | |
|---|---|---|---|
| Components | Weight Fraction [%] | Components | Molar Fraction |
| $SiO_2$<br>$Al_2O_3$ | 55<br>39 | $H_2O$ | 73.7 |
| $K_2O + Na_2O$<br>$Fe_2O_3$ | 1<br>1.8 | KOH | 11.7 |
| $TiO_2$<br>$CaO + MgO$ | 1.5<br>0.6 | $SiO_2$ | 14.6 |

## 2.2. APM and ZPM Solutions

The APM solution is made by mixing of alumina powder, particle size 0.5 μm, with the activator defined in Table 1, mass ratio between alumina powder and activator being 1/1. The ZPM solution is made by mixing of a zirconia powder, particle size 0.4 μm, with the activator defined in Table 1, mass ratio between Zr powder and activator being 1/1.

The composite coatings result in a paste-like solution, and consolidation reactions then follow because of dehydration of the APM and ZPM solutions. Mechanical strength results from the chemical bonds between the basic potassium silicate in alkali solution chains and the acid alumina and zirconia powders.

## 2.3. Specimens Preparation

Eight composite components are obtained by applying coating layers on the two opposite faces of the beam substrate. The specimen geometry specifications are length $(11.0 \pm 0.01) \times 10^{-3}$ m, thickness $(0.5 \pm 0.01) \times 10^{-3}$ m and width $(3.0 \pm 0.01) \times 10^{-3}$ m.

Specimen data are reported in Table 2, including the specimen label, the substrate and the coating layer material, the production technique, and the coating layer thickness. Table 1 label "raw" indicates that the surface substrate is unfinished, texture $R_a$ 0.8, while "sdb" label refers to a sandblasted surface substrate, texture $R_a$ 12. After applying the GP, APM and ZPM coatings to the metal substrates, all samples were cured at $T = 25\ °C$ for 9 days, in order to increase the adhesion behaviour [35].

**Table 2.** Specimen data.

| Specimen | Substrate | Coating Layer |
|----------|-----------|---------------|
| A1 | Al1000 (raw) | GP |
| A2 | Al1000 (sdb) | GP |
| A3 | Al1000 (raw) | APM |
| A4 | Al1000 (raw) | ZPM |
| S1 | Steel C67 (raw) | GP |
| S2 | Steel C67 (sdb) | GP |
| S3 | Steel C67 (raw) | APM |
| S4 | Steel C67 (raw) | ZPM |

Note: the production technique is screen printing and the coating thickness is 125 μm.

According to known literature [35,36], a geopolymer cured at room temperature tends to slowly change its structure and presents a low porosity and a high toughness while when cured at a higher temperature it exhibits faster structure changes, higher porosity and lower toughness.

In previous works [22,23], $TiO_2$ and $Al_2O_3$ based coatings were considered, but the results did not show a meaningful improvement with respect to the uncoated beam specimen, concerning both vibration damping behaviour and adhesion strength in cyclic loading condition. In this contribution, new layer technologies based on a screen printing production process and inorganic polymer based composite material solutions, are taken into account.

### 2.4. Optical and SEM Specimen Structural Characterization

The coating structure and the coating-substrate interface is analyzed by means of optical and scanning electron microscopy (SEM). Samples were fracture cut in the transverse cross section, in order to expose the whole cross section. Before optical investigation, the specimen cross sections were embedded in epoxy resin and polished with abrasive SiC paper up to 2500 mesh and then by using a diamond based, 0.5 μm particle size, lapping paste. Chemical etching (Ethanol added to 3% $HNO_3$ at 150 °C) then follows.

Figure 1a–c report the SEM images related to the three coatings solutions proposed. All coatings present a typical composite structure, showing: geopolymer and unreacted potassium silicate (Figure 1a), the potassium silicate as matrix and fine alumina dispersed particles (Figure 1b), the potassium silicate as matrix and zirconia dispersed particles (Figure 1c). In Figure 1b the 0.5 μm particle size is observed and the almost total absence of shrinkage and solidification defects to form rigid structures, possibly associated to big strength characteristics, is also outlined. In Figure 1c, the size of zirconia powder is difficult to evaluate because of the resulting irregular morphology with brittle fragments of consolidate and dehydrate potassium silicate.

(a)  (b)  (c)

**Figure 1.** SEM images of the different coating solutions fracture structure: (**a**) GP; (**b**) APM; (**c**) ZPM.

Figure 1a, referring to the geopolymeric coating, shows a compacted interconnected microstructure which can increase toughness and strength in comparison to the other coatings considered. The amorphous nature of these coatings may greatly influence the stress state of consolidate coatings, so that there are no detectable cracks and microcracks that could compromise the performance of the coatings when subjected to mechanical strain vibrations, because no differences induced by thermal and mechanical stresses into the polycrystalline state are expected to appear in working conditions.

Figure 2a–f report the optical images related to the interface obtained by means of the three coatings solutions and two metal substrates investigated. As shown, there is an evident adhesion between ceramic coatings and the metal substrate.

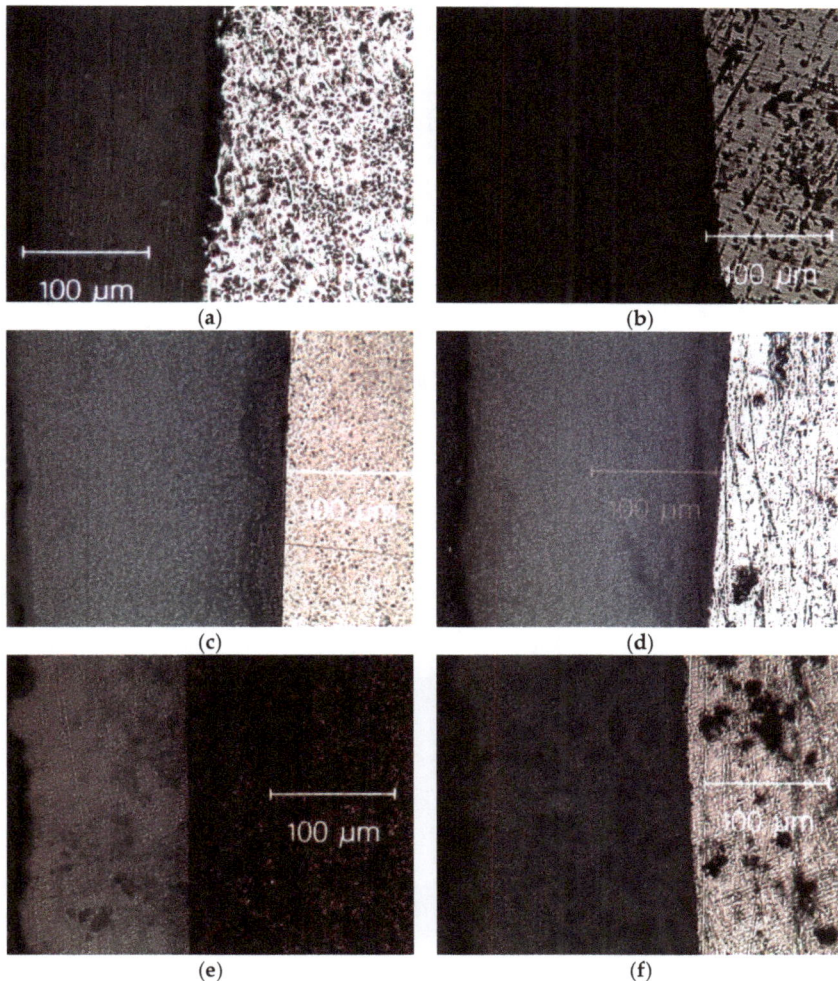

**Figure 2.** Optical micrograph images (200×) of the interface between coating (left) and metal substrate (right): (**a**) GP/C67; (**b**) GP/Al1000; (**c**) APM/C67; (**d**) APM/Al1000; (**e**) ZPM/C67; (**f**) ZPM/Al1000.

Figure 3a–f report the SEM images related to the interface obtained by means of the three coatings solutions and two metal substrates investigated. Defects by shrinkage phenomena at the metal coating interface are not shown. It can be outlined that the adhesion between the proposed coating solutions and the Al1000 substrate appears to be more effective than with respect to the C67 substrate.

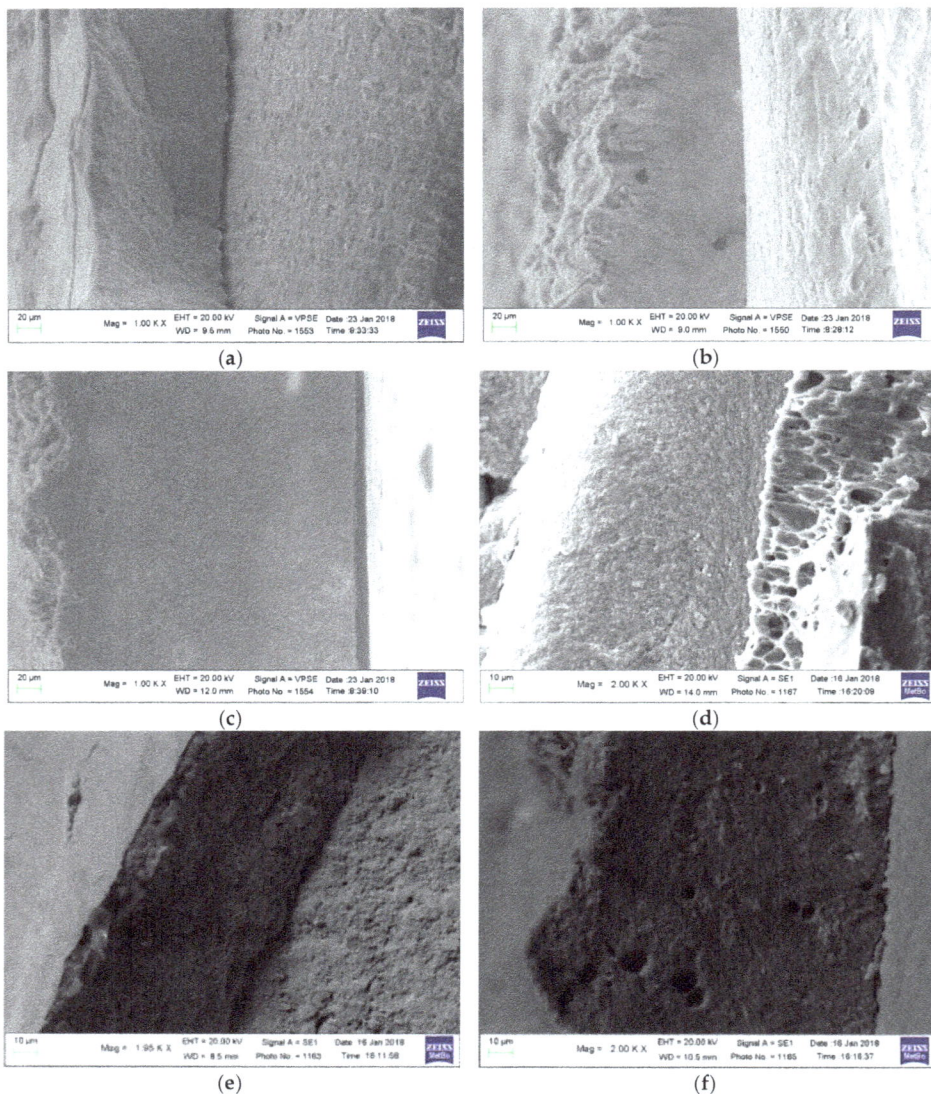

**Figure 3.** SEM images of the of the interface between coating (left) and metal substrate (right): (**a**) GP/C67; (**b**) GP/Al1000; (**c**) APM/C67; (**d**) APM/Al1000; (**e**) ZPM/C67; (**f**) ZPM/Al1000.

## 3. Identification Procedure

In this work, applied force and displacement data obtained by means of dynamic mechanical measurements over a wide frequency range are used to estimate the complex modulus $E(\omega)$ of measured beam specimens. The values of $E(j\omega)$ are estimated by means of a procedure fully defined by the authors of this work in a previous paper [29] and are here briefly outlined.

$E(j\omega)$ defines the specimen stress ($\hat{\sigma}$)-strain ($\hat{\varepsilon}$) equivalent material relationship in the frequency domain (Equation (1)):

$$\hat{\sigma} = E(j \times \omega) \times \hat{\varepsilon} \tag{1}$$

Slender beams of uniform rectangular cross section with clamped sliding boundary conditions are considered. The contribution of the inertial actions is also taken into account.

The complex, experimentally estimated values of $E_i = E(\omega_i)$ at frequencies $\omega_i$ are found by finding the solutions of Equation (2) by means of the Newton Raphson method, starting from known static modulus $E_0 = E(j\omega = 0)$:

$$\Theta(E_i) = \sum_{s=1}^{n_m} \frac{\left[\frac{2 \times \sinh k_s \times \sin k_s}{\cosh k_s + \cos k_s}\right]^2}{E(j \times \omega_i) \times k_s^4 \times I/L^3 - M \times \omega_i^2} - \frac{\hat{v}(j \times \omega_i)}{\hat{q}(j \times \omega_i)} = 0 \tag{2}$$

$\hat{v}(j\omega_i)$ and $\hat{q}(j\omega_i)$ are respectively the measured transverse displacement and applied periodic force (at frequency $\omega_i$) at the beam sliding end. $M$ is the beam mass, $I$ is the beam section moment, $L$ is the beam length and $k_s$ are the roots of $f(k_s) = \tan(k_s) + \tanh(k_s) = 0$.

A specific robust identification and condensation procedure [37] is applied on the $E(j\omega)$ experimentally estimated values to identify the specimens stress-strain relationship in the frequency domain. $E(j\omega)$ (Equation (1)) is modeled by means of a high order generalized Kelvin model and its parameters are identified. The global model order can be condensed to obtain a new model of comparable accuracy but significantly lower order.

## 4. Experimental Set-Up, Measurement and Discussion

### 4.1. Experimental Set-Up

The dynamic mechanical tests are realized with a standard dynamic mechanical analyzer apparatus (TA Instrument DMA Q800, New Castle, DE, USA) in a forced flexural excitation, harmonic sine, experimental set up Figure 4, clamped sliding boundary conditions, 0.01% maximum strain, 0.01–200 Hz frequency range with minimal frequency resolution $\Delta f = 0.01$ Hz and air flow 35 °C isothermal conditions. Transverse displacement and applied periodic force amplitude, at frequency $\omega_i$, are measured at the beam sliding end.

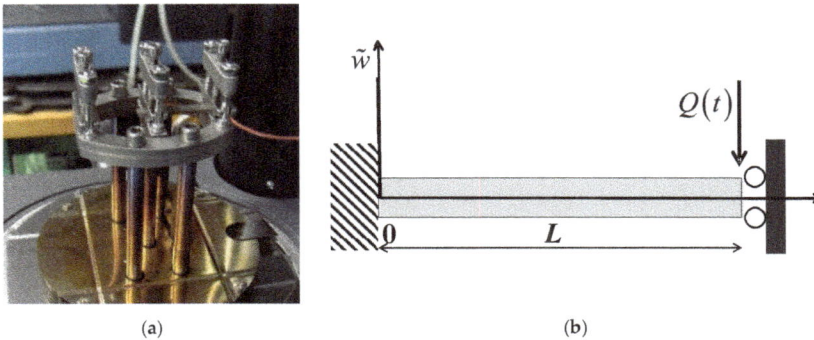

(a)                                                            (b)

**Figure 4.** (**a**) DMA single-cantilever flexural experimental set up with specimen; (**b**) Schematic representation of a slender beam in the single-cantilever experimental set-up.

The experimentally $E(j\omega)$ estimated values are obtained by means of the procedure reported in Section 3. The specimen dissipative properties are estimated by $z(j\omega) = \text{Im}(E)/|E|$, a normalized real coefficient belonging to the [0,1] range, so that being it different with respect to the standard approach based on $\tan\delta = \text{Im}(E)/\text{Re}(E)$ choice [38] and meaningful when used to compare different solutions. The results obtained for the homogeneous and the composite specimens are compared and shown in Figures 5–8.

## 4.2. Experimental Results and Discussion

The results obtained for specimens coated with GP (A1, A2, S1 and S2), reported in Figures 5 and 6, show a general increase of the damping properties with respect to the uncoated specimens. The damping increase appears to depend on the type of the substrate and the surface texturing.

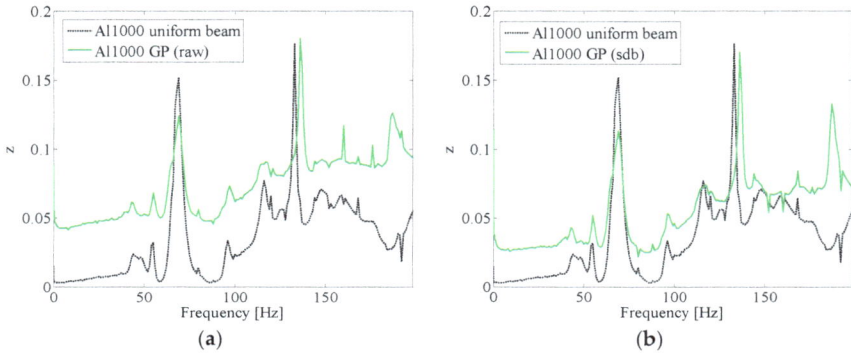

**Figure 5.** $z(j\omega)$ estimates for A1 (**a**) and A2 (**b**) specimen and homogeneous Al1000 specimen.

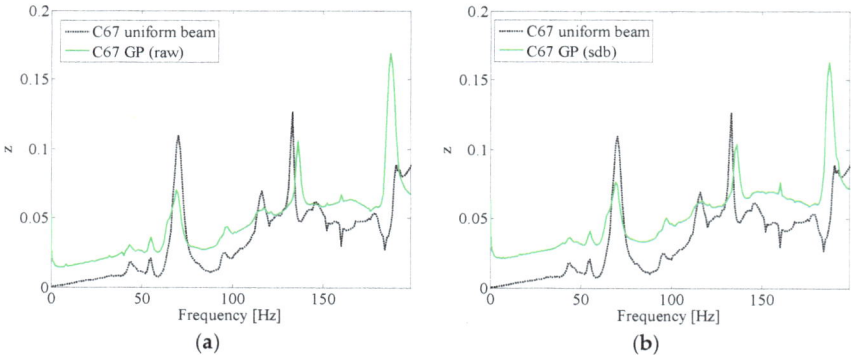

**Figure 6.** $z(j\omega)$ estimates for S1 (**a**) and S2 (**b**) specimen and homogeneous C67 specimen.

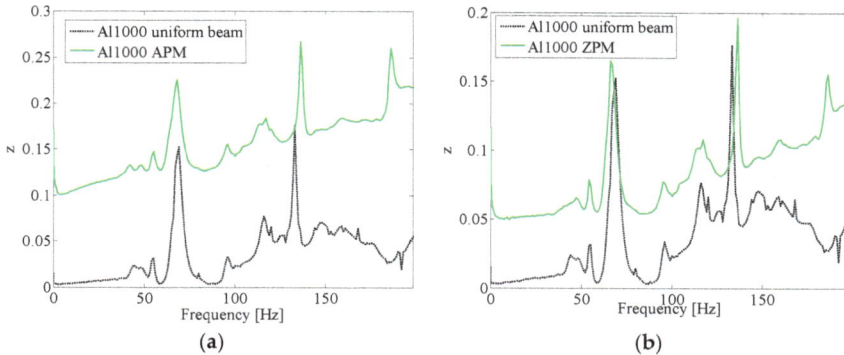

**Figure 7.** $z(j\omega)$ estimates for A3 (**a**) and A4 (**b**) specimen and homogeneous Al1000 specimen.

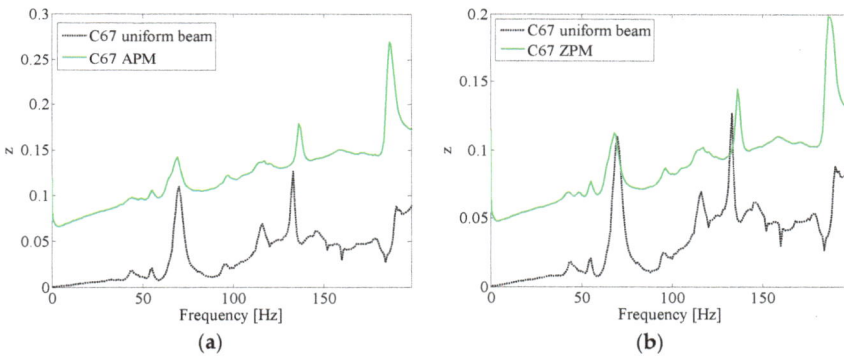

**Figure 8.** $z(j\omega)$ estimates for S3 (**a**) and S4 (**b**) specimen and homogeneous C67 specimen.

The A1 specimen shows the largest $z(j\omega)$ increase, A2 and S2 specimens show a similar increase of $z(\omega)$, while S1 component shows to be the less effective solution of this set. The results obtained from specimens A3, A4, S3 and S4 (APM, ZPM) are plotted in Figures 7 and 8, showing the greatest improvement in damping behavior. As in the previous case, the effectiveness of the coating solution is also dependent on the substrate material, since both the APM (A3, S3) and ZPM (A4, S4) composite solutions appear to be more effective on Al1000 than on C67 substrate.

Coating solutions adopted in specimens A1–4 and S1–4 can be considered more effective than other solutions previously investigated by the authors of this work [22,23] and other researchers in the same field (in the authors' knowledge).

Figure 9 shows the $z(j\omega)$ ratio estimate with respect to homogeneous specimens made with the GP, APM and ZPM coating materials. The three coating materials exhibit good damping capabilities, with the GP specimen displaying the lower $z(j\omega)$ ratio and the ZPM specimen being associated to the most effective solution. Figures 5–8 indicate that the damping behaviour is strongly influenced by the dissipative actions (friction) exhibited at the interface between composite substrate and the coating layer, thus dominating the effect of the internal dissipative actions of the coating material. In Figures 5 and 6, same substrate and coating layers but different substrate surface texture influences the damping behaviour, while in Figures 7 and 8, while there is the same coating and surface texture, the damping behaviour is influenced by the substrate material. It appears that A3 is the most effective solution of the evaluated set, and the APM coating is less effective than ZPM from the inherent material standpoint (Figure 9), so that enforcing the assumption that dissipative actions mainly depend on the interlaminar interface structure and do not depend on the distributed material coating properties. Figure 10a shows a condensed global model of order $n = 13$ for the A3 specimen and Figure 10b shows a condensed global model of order $n = 14$ for the S3 specimen. In both cases an identified global model of significantly higher order ($n = 43$) is initially obtained and then processed by means of the condensation procedure and of the evaluation and processing of pole stability diagrams.

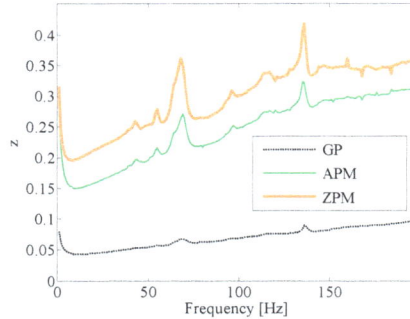

**Figure 9.** $z(j\omega)$ estimates for homogeneous specimens of GP, APM and ZPM.

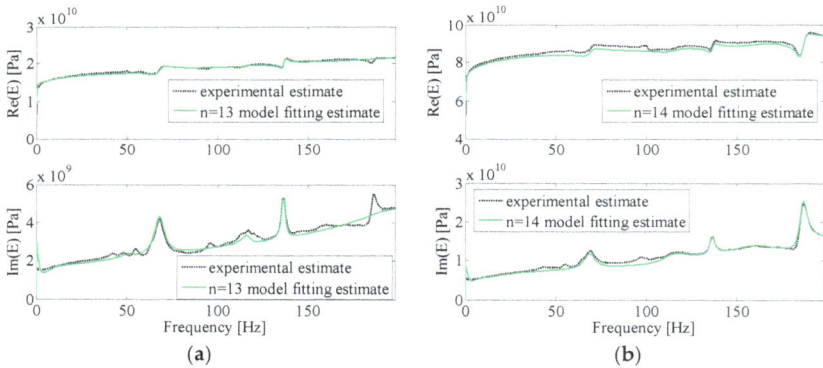

**Figure 10.** (**a**) A4 specimen $E(\omega)$ experimental estimate versus $n = 13$ condensed model fitting estimate; (**b**) S4 specimen $E(\omega)$ experimental estimate versus $n = 14$ condensed model fitting estimate.

## 5. Modeling of Multi-Layer Coated Beam Composites

### 5.1. Motivations

The contribution of interface dissipative actions to the damping behaviour of coated components is clearly outlined from the results presented in the previous sections. Nevertheless, while the procedure used to obtain the condensed, generalized standard linear solid model $E(j\omega)$ may be indeed effective to estimate the damping properties of different specimens and to compare them, it cannot be used to design and predict the damping behaviour of new, not previously tested, coating solutions. An effective model must be able to take into account the coating solution architecture, i.e., the number of layers, the thickness, the material and the coating technology adopted for the different layers as well.

In literature many models are known for multi-layered beams and plates [29,39–44] but no attempts can be found with respect to modelling the interface dissipative actions. In this section, a multi-layered beam model based on zig-zag beam and plates theory addressing this issue is presented. The model takes into account the layers of geometric and material properties and is able to deal with interface slipping and local friction. The model uses an elasto-hysteretic contribution to define the dissipative actions at the interface between the layers.

*5.2. Multi-Layer Beam Flexural Model*

A schematic representation of a multi-layered beam model is reported in Figure 11. A multi-layered beam made up of $N$ layers and with uniform rectangular section is considered. $L, g, h$, are respectively the beam length, width and thickness, $V = L \times g \times h$, $\rho_i$ is the $i$-th layer density. In each layer ($i = 1, \ldots, N$), the material constitutive equations are assumed to be:

$$\sigma(\lambda \in [\lambda_{i-1}, \lambda_i]) = E_i \times \varepsilon, \tau(\lambda \in [\lambda_{i-1}, \lambda_i]) = G_i \times \gamma, i = 1, \ldots, N \qquad (3)$$

where $E_i$ and $G_i$ are the $i$-th layer longitudinal and shear elastic moduli, $\sigma = \sigma_{xx}$ and $\tau = \tau_{xy}$ are the flexural stress and shear stress components, $\varepsilon = \varepsilon_{xx}$ and $\gamma = \gamma_{xy}$ are the strain components. Since small displacement and deformation fields are considered, transverse displacement $\tilde{w}$ is assumed as being stationary with respect to $y, \lambda$. Transverse and longitudinal displacement $\tilde{w}, \tilde{u}$ are assumed as follows:

$$\begin{aligned}
&\tilde{w} = L \times w(\chi, t), \; \tilde{u} = h \times u(\chi, \lambda, t) \\
&u(\chi, \lambda, t) = a + b \times \lambda + c \times \lambda^2 + d \times \lambda^3, 0 \le \lambda \le \lambda_1 \\
&u(\chi, \lambda, t) = a + b \times \lambda + c \times \lambda^2 + d \times \lambda^3 + (a_i + b_i \times \lambda), \lambda_{i-1} \le \lambda \le \lambda_i, i = 2 \ldots N
\end{aligned} \qquad (4)$$

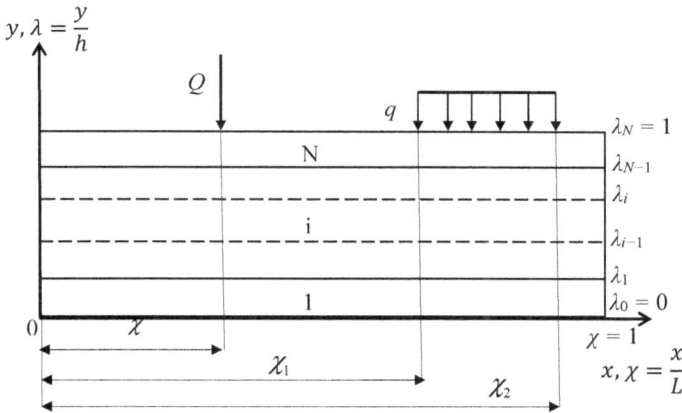

**Figure 11.** Multi-layer beam model schematic representation.

The following $2 \times N + 3$ state variables result:

$$\left\{ \begin{array}{cccccccccc} w(\chi, t) & a(\chi, t) & b(\chi, t) & c(\chi, t) & d(\chi, t) & a_2(\chi, t) & b_2(\chi, t) & \cdots & a_N(\chi, t) & b_N(\chi, t) \end{array} \right\}^T \qquad (5)$$

The kinematical relationships between strain and displacement components is assumed to be:

$$\begin{aligned}
&\varepsilon(\chi, \lambda, t) = \tfrac{h}{L} \times u' \\
&\gamma(\chi, \lambda, t) = w' + b + 2 \times c \times \lambda + 3 \times d \times \lambda^2, \quad \lambda \le \lambda_1 \\
&\gamma(\chi, \lambda, t) = \tfrac{\partial u}{\partial \lambda} + w' = w' + b + b_i + 2 \times c \times \lambda + 3 \times d \times \lambda^2, \quad \lambda_{i-1} \le \lambda \le \lambda_i
\end{aligned} \qquad (6)$$

where $(\ )' = \partial(\ )/\partial\chi = L \times \partial(\ )/\partial x$.

$N + 1$ equilibrium conditions at the layer interfaces hold:

$$\tau(0) = G_1 \times \gamma(0) = 0, \ \tau(1) = G_N \times \gamma(1) = 0 \tag{7}$$

$$\tau\left((\lambda_i)^-\right) = G_i \times \gamma\left((\lambda_i)^-\right) = G_{i+1} \times \gamma\left((\lambda_i)^+\right) = \tau\left((\lambda_i)^+\right), \quad i = 1, \ldots, N-1 \tag{8}$$

where $(\ )^- = \lim\limits_{\Delta \to 0}((\ ) - \Delta), \ (\ )^+ = \lim\limits_{\Delta \to 0}((\ ) + \Delta)$

From Equation (7):

$$b = -w', b_N = -2 \times c - 3 \times d \tag{9}$$

From Equations (6), (8) and (9), a system of $N - 1$ equations can be obtained:

$$\begin{cases} G_2 \times b_2 = \left(3 \times d \times \lambda_1^2 + 2 \times c \times \lambda_1\right) \times (G_1 - G_2) \\ \ldots \\ G_N \times b_N - G_k \times b_{N-1} = \left(3 \times d \times \lambda_{N-1}^2 + 2 \times c \times \lambda_N\right) \times (G_{N-1} - G_N) \end{cases} \tag{10}$$

By equating the Equation (10) right side sum to Equation (10) left side sum, and taking into account of Equation (9):

$$c = \bar{c} \times d, \quad \bar{c} = -\frac{3}{2} \times \bar{\lambda}, \quad \bar{\lambda} = \frac{\left\{ \sum\limits_{i=1}^{N-1} \left[\lambda_i^2 \times (G_i - G_{i+1})\right] + G_N \right\}}{\left\{ \sum\limits_{i=1}^{N-1} \left[\lambda_i \times (G_i - G_{i+1})\right] + G_N \right\}} \tag{11}$$

$c$ is a stationary value defined by the shear moduli and thickness values of the beam layers. By defining the following variable change:

$$b_i = \bar{b}_i \times d, \quad i = 2 \ldots N \tag{12}$$

Putting Equations (10)–(12), the following iterative formula results:

$$\begin{aligned} \bar{b}_2 &= 3 \times \lambda_1 \times \frac{G_1 - G_2}{G_2} \times \left(\lambda_1 - \bar{\lambda}\right) \\ \bar{b}_{i+1} &= \frac{G_i}{G_{i+1}} \times \bar{b}_i + 3 \times \lambda_i \times \frac{G_i - G_{i+1}}{G_{i+1}} \times \left(\lambda_i - \bar{\lambda}\right) \qquad i = 2 \ldots N-1 \end{aligned} \tag{13}$$

Now by relaxing the continuity of the kinematical $u$ component at $\lambda = \lambda_i$ interface, the longitudinal sliding $v(\lambda) = u\left((\lambda)^+\right) - u\left((\lambda)^-\right)$ at the interface is:

$$\begin{aligned} v(\lambda_1, \chi, t) &= v_1(\chi, t) = \left(\bar{b}_2 \times \lambda_1 \times d + a_2\right) \\ v(\lambda_i, \chi, t) &= v_i(\chi, t) = \left(\left(\bar{b}_{i+1} - \bar{b}_i\right) \times \lambda_i \times d + a_{i+1} - a_i\right), \qquad i = 2, \ldots, N-1 \end{aligned} \tag{14}$$

At $\lambda = \lambda_i$ interface, an elasto-hysteretic constitutive relationship is assumed by means of complex impedance $\Phi_i$:

$$\tau(\lambda_i) = \Phi_i \times h \times v_i(\chi, t), \Phi_i = \varphi_i + j \times \eta_i, \quad \Omega_i = \Phi_i^{-1} = \frac{\varphi_i - j \times \eta_i}{|\Phi_i|^2}; \quad i = 1, \ldots, N-1 \tag{15}$$

where $j = \sqrt{-1}$. Defining the following variable change:

$$a_i = \bar{a}_i \times d, \qquad i = 2 \ldots N \tag{16}$$

the values of $a_i$ from the following iterative formula, obtained by combining Equations (3), (14) and (15):

$$\begin{aligned}
\bar{a}_2 &= -3 \times (\lambda_1 - \overline{\lambda}) \times \lambda_1 \times \left( \frac{\Omega_1 \times G_1}{h} + \frac{G_1 - G_2}{G_2} \right) \\
\bar{a}_{i+1} &= \bar{a}_i - \left( \bar{b}_{i+1} - \bar{b}_i \right) \times \lambda_i + \frac{\Omega_i \times G_i}{h} \times \left( 3 \times (\lambda_i - \overline{\lambda}) \times \lambda_i + \bar{b}_i \right), \quad i = 2, \ldots, N-1
\end{aligned} \tag{17}$$

From Equations (11), (13) and (17), stationary $\bar{a}_i, \bar{b}_i, \bar{c}$, values result. Only three independent state variables, collected in vector **X**, result:

$$\mathbf{X} = \mathbf{X}(\chi, t) = \left\{ \begin{array}{ccc} w(\chi, t) & a(\chi, t) & d(\chi, t) \end{array} \right\}^T \tag{18}$$

The following expression of the kinematical components result:

$$w = \begin{bmatrix} 0 & 1 & 0 \end{bmatrix} \times \mathbf{X}, \quad u = \begin{bmatrix} 1 & -\lambda & \theta(\lambda) \end{bmatrix} \times \begin{bmatrix} 1 & 0 & 0 \\ 0 & ()' & 0 \\ 0 & 0 & 1 \end{bmatrix} \times \mathbf{X}$$

$$\varepsilon = \frac{h}{L} \times \begin{bmatrix} 1 & -\lambda & \theta(\lambda) \end{bmatrix} \times \begin{bmatrix} ()' & 0 & 0 \\ 0 & ()'' & 0 \\ 0 & 0 & ()' \end{bmatrix} \times \mathbf{X}, \quad \gamma = \frac{\partial \theta(\lambda)}{\partial \lambda} \times \begin{bmatrix} 0 & 0 & 0 \\ 0 & 0 & 0 \\ 0 & 0 & 1 \end{bmatrix} \times \mathbf{X} \tag{19}$$

$$\theta(\lambda \le \lambda_1) = \left( \lambda - \frac{3}{2} \times \overline{\lambda} \right) \times \lambda^2$$

$$\theta(\lambda_{k-1} \le \lambda \le \lambda_k) = \left( \lambda - \frac{3}{2} \times \overline{\lambda} \right) \times \lambda^2 + \bar{b}_i \times \lambda + \bar{a}_i$$

### 5.3. Equation of Motion

To obtain the equation of motion, the system total potential energy ($\Pi$) is considered:

$$\Pi = U + W_I + W_E + \Delta\Pi \tag{20}$$

$U$ is the contribution of the internal elasto-hysteretic actions:

$$U = \frac{V}{2} \times \left( \int_0^1 \int_0^1 \sigma \times \varepsilon \times d\lambda \times d\chi + \int_0^1 \int_0^1 \tau \times \gamma \times d\lambda \times d\chi \right) \tag{21}$$

$W_I$ is the contribution of the inertial actions:

$$W_I = V \times \left( h^2 \times \int_0^1 \int_0^1 \rho \times u \times \ddot{u} \times d\lambda \times d\chi + L^2 \times \int_0^1 \int_0^1 \rho \times w \times \ddot{w} \times d\lambda \times d\chi \right) \tag{22}$$

where $(\ddot{\ }) = \partial^2(\ )/\partial t^2$. $W_E$ is the contribution of the external forces:

$$W_E = -Q \times L \times w(\overline{\chi}) - L^2 \times g \times \int_{\chi_1}^{\chi_2} q(\chi) \times w \times d\chi \tag{23}$$

where $Q$ is a lumped force applied at $\overline{\chi}$ and $q$ is a distributed pressure applied at $\chi_1 \le \chi \le \chi_2$.

$\Delta\Pi$ is the contribution of the distributed, viscous elastic, constraints modeling boundary conditions:

$$\Delta\Pi = \Delta\Pi_e + \Delta\Pi_v$$

$$\Delta\Pi_e = \frac{V}{2} \times \left( L^2 \times \int_0^1 K_w(\chi) \times w \times w \times d\chi + h^2 \times \int_0^1 \int_0^1 K_u(\chi,\lambda) \times u \times u \times d\lambda \times d\chi \right) \tag{24}$$

$$\Delta\Pi_v = V \times \left( L^2 \times \int_0^1 C_w(\chi) \times w \times \dot{w} \times d\chi + h^2 \times \int_0^1 \int_0^1 C_u(\chi,\lambda) \times u \times \dot{u} \times d\lambda \times d\chi \right)$$

where $(\dot{\ }) = \partial(\ )/\partial t$ and $K_w, K_u, C_w, C_u$, are the constraint elastic and viscous parameters respectively. It is assumed that the state unknown solution variables of Equation (18) satisfy:

$$X_r(\chi,t) = \alpha_r(\chi) \times \delta_r(t), \quad r = 1,2,3 \tag{25}$$

and generally unknown functions $\alpha_r$ are restricted to a set of known harmonic functions:

$$\mathbf{X}(\chi,t) = \left\{ \begin{array}{ccc} a & w & d \end{array} \right\}^T = \mathbf{\Psi}(\chi) \times \delta(t) = \begin{bmatrix} \mathbf{\Psi}_a & 0 & 0 \\ 0 & \mathbf{\Psi}_w & 0 \\ 0 & 0 & \mathbf{\Psi}_d \end{bmatrix} \times \delta \tag{26}$$

$$\begin{array}{l} \mathbf{\Psi}_a = \sqrt{2} \times \left[ \begin{array}{cccccc} \frac{1}{\sqrt{2}} & \sin(\pi \times \chi) & \cos(\pi \times \chi) & \ldots & \sin(n_a \times \pi \times \chi) & \cos(n_a \times \pi \times \chi) \end{array} \right] \\ \mathbf{\Psi}_w = \sqrt{2} \times \left[ \begin{array}{cccccc} \frac{1}{\sqrt{2}} & \sqrt{12} \times (\chi - 0.5) & \sin(\pi \times \chi) & \cos(\pi \times \chi) & \ldots & \sin(n_w \times \pi \times \chi) & \cos(n_w \times \pi \times \chi) \end{array} \right] \\ \mathbf{\Psi}_d = \sqrt{2} \times \left[ \begin{array}{ccccc} \sin(\pi \times \chi) & \cos(\pi \times \chi) & \ldots & \sin(n_d \times \pi \times \chi) & \cos(n_d \times \pi \times \chi) \end{array} \right] \\ \delta = \left[ \begin{array}{ccc} \delta_1 & \ldots & \delta_n \end{array} \right]^T, \quad n = 2 \times (n_a + n_w + n_d) + 3 \end{array} \tag{27}$$

And using Equations (23) and (27):

$$w = \left[ \begin{array}{ccc} 0 & \mathbf{\Psi}_w & 0 \end{array} \right] \times \delta, \quad u = \mathbf{B}^T(\lambda) \times \mathbf{\Psi_1} \times \delta,$$

$$\varepsilon = \frac{h}{L} \times \mathbf{B}^T(\lambda) \times \mathbf{\Psi_2} \times \delta, \quad \gamma = \frac{\partial\theta(\lambda)}{\partial\lambda} \times \begin{bmatrix} 0 & 0 & 0 \\ 0 & 0 & 0 \\ 0 & 0 & \mathbf{\Psi}_d \end{bmatrix} \times \delta \tag{28}$$

where:

$$\mathbf{B}(\lambda) = \begin{bmatrix} 1 \\ -\lambda \\ \theta(\lambda) \end{bmatrix}, \quad \mathbf{\Psi_1} = \begin{bmatrix} \mathbf{\Psi}_a & 0 & 0 \\ 0 & \mathbf{\Psi}'_w & 0 \\ 0 & 0 & \mathbf{\Psi}_d \end{bmatrix}, \quad \mathbf{\Psi_2} = \begin{bmatrix} \mathbf{\Psi}'_a & 0 & 0 \\ 0 & \mathbf{\Psi}''_w & 0 \\ 0 & 0 & \mathbf{\Psi}'_d \end{bmatrix} \tag{29}$$

Using Equations (26)–(29), from Equations (21)–(24) results:

$$U = \frac{1}{2} \times \delta^T \times (\mathbf{K}_\varepsilon + \mathbf{K}_\gamma) \times \delta,$$

$$\mathbf{K}_\varepsilon = \frac{V \times h^2}{L^2} \times \int_0^1 \mathbf{\Psi_2}^T \times \mathbf{B}_E \times \mathbf{\Psi_2} \times d\chi, \quad \mathbf{K}_\gamma = V \times \sum_{i=1}^N G_i \times \int_{\lambda_{i-1}}^{\lambda_i} \left( \frac{\partial\theta(\lambda)}{\partial\lambda} \right)^2 \times d\lambda \times \begin{bmatrix} 0 & 0 & 0 \\ 0 & 0 & 0 \\ 0 & 0 & \tilde{\mathbf{\Psi}}_d \end{bmatrix},$$

$$\mathbf{B}_E = \sum_{i=1}^N E_i \times \int_{\lambda_{i-1}}^{\lambda_i} \mathbf{B}(\lambda) \times \mathbf{B}^T(\lambda) \times d\lambda, \quad \tilde{\mathbf{\Psi}}_d = \int_0^1 \mathbf{\Psi}_d^T(\chi) \times \mathbf{\Psi}_d(\chi) \times d\chi \tag{30}$$

$$W_I = \delta^T \times (\mathbf{M}_u + \mathbf{M}_w) \times \ddot{\delta},$$

$$\mathbf{M}_u = V \times h^2 \times \int_0^1 \mathbf{\Psi_1}^T \times \mathbf{B}_\rho \times \mathbf{\Psi_1} \times d\chi, \quad \mathbf{M}_w = V \times L^2 \times \left( \sum_{i=1}^N \rho_i \times (\lambda_i - \lambda_{i-1}) \right) \times \begin{bmatrix} 0 & 0 & 0 \\ 0 & \tilde{\mathbf{\Psi}}_w & 0 \\ 0 & 0 & 0 \end{bmatrix} \tag{31}$$

$$\mathbf{B}_\rho = \sum_{i=1}^N \rho_i \times \int_{\lambda_{i-1}}^{\lambda_i} \mathbf{B}(\lambda) \times \mathbf{B}^T(\lambda) \times d\lambda, \quad \tilde{\mathbf{\Psi}}_w = \int_0^1 \mathbf{\Psi}_w^T(\chi) \times \mathbf{\Psi}_w(\chi) \times d\chi$$

$$W_E = -\delta^T \times \mathbf{Q}, \quad \mathbf{Q} = L \times \left( Q \times \left\{ \begin{matrix} 0 \\ \mathbf{\Psi}_w^T(\overline{\chi}) \\ 0 \end{matrix} \right\} + L \times g \times \int\limits_{\chi_1}^{\chi_2} q(\chi) \times \left\{ \begin{matrix} 0 \\ \mathbf{\Psi}_w^T(\chi) \\ 0 \end{matrix} \right\} \times d\chi \right) \tag{32}$$

$$\Delta\Pi_e = \tfrac{1}{2} \times \delta^T \times (\Delta \mathbf{K}_W + \Delta \mathbf{K}_u) \times \delta$$

$$\Delta \mathbf{K}_W = \begin{bmatrix} 0 & 0 & 0 \\ 0 & V \times L^2 \times \widetilde{\mathbf{\Psi}}_e & 0 \\ 0 & 0 & 0 \end{bmatrix}, \quad \Delta \mathbf{K}_u = V \times h^2 \times \int\limits_0^1 \mathbf{\Psi_1}^T(\chi) \times \mathbf{B}_e \times \mathbf{\Psi_1}(\chi) \times d\chi \tag{33}$$

$$\mathbf{B}_e = \sum_{i=1}^N K_u(\chi) \times \int\limits_{\lambda_{i-1}}^{\lambda_i} \mathbf{B}(\lambda) \times \mathbf{B}^T(\lambda) \times d\lambda, \quad \widetilde{\mathbf{\Psi}}_e = \int_0^1 K_w(\chi) \times \mathbf{\Psi}_w^T(\chi) \times \mathbf{\Psi}_w(\chi) \times d\chi$$

$$\Delta\Pi_v = \delta^T \times (\Delta \mathbf{C}_w + \Delta \mathbf{C}_u) \times \dot{\delta}$$

$$\Delta \mathbf{C}_w = \begin{bmatrix} 0 & 0 & 0 \\ 0 & V \times L^2 \times \widetilde{\mathbf{\Psi}}_v & 0 \\ 0 & 0 & 0 \end{bmatrix}, \quad \Delta \mathbf{K}_u = V \times h^2 \times \int\limits_0^1 \mathbf{\Psi_1}^T(\chi) \times \mathbf{B}_v \times \mathbf{\Psi_1}(\chi) \times d\chi \tag{34}$$

$$\mathbf{B}_v = \sum_{i=1}^N C_u(\chi) \times \int\limits_{\lambda_{i-1}}^{\lambda_i} \mathbf{B}(\lambda) \times \mathbf{B}^T(\lambda) \times d\lambda, \quad \widetilde{\mathbf{\Psi}}_v = \int_0^1 C_w(\chi) \times \mathbf{\Psi}_w^T(\chi) \times \mathbf{\Psi}_w(\chi) \times d\chi$$

and the resulting equation of motion is:

$$(\mathbf{M}_u + \mathbf{M}_w) \times \ddot{\delta} + (\Delta\mathbf{C}_w + \Delta\mathbf{C}_u) \times \dot{\delta} + [(\mathbf{K}_\varepsilon + \mathbf{K}_\gamma) + (\Delta\mathbf{K}_w + \Delta\mathbf{K}_u)] \times \delta = \mathbf{Q}. \tag{35}$$

Because of the elasto-hysteretic assumption (Equation (15)), complex $\mathbf{M}_u$, $\Delta\mathbf{C}_u$, $\mathbf{K}_\varepsilon$, $\Delta\mathbf{K}_u$ matrices result. To find the beam frequency response function $\widetilde{F}(\chi_w, \chi_q, j\omega)$, i.e., the complex transverse response $\widetilde{w}(\chi_w, j\omega)$ related to unitary, lumped, harmonic excitation at frequency $\omega$ applied at abscissa $\chi_q$, the following procedure is applied. Both sides of Equation (35) are Fourier transformed $\mathcal{F}() = (\hat{})$

$$\left[ -\omega^2 \times (\mathbf{M}_u + \mathbf{M}_w) + j \times \omega \times (\Delta\mathbf{C}_w + \Delta\mathbf{C}_u) + (\mathbf{K}_\varepsilon + \mathbf{K}_\gamma) + (\Delta\mathbf{K}_w + \Delta\mathbf{K}_u) \right] \times \hat{\delta} = \hat{\mathbf{Q}}. \tag{36}$$

The frequency response function matrix $\mathbf{F}(j\omega)$ is:

$$\mathbf{F}(j \times \omega) = \left[ -\omega^2 \times (\mathbf{M}_u + \mathbf{M}_w) + j \times \omega \times (\Delta\mathbf{C}_w + \Delta\mathbf{C}_u) + (\mathbf{K}_\varepsilon + \mathbf{K}_\gamma) + (\Delta\mathbf{K}_w + \Delta\mathbf{K}_u) \right]^{-1}. \tag{37}$$

Equation (37) can be expressed as:

$$\mathbf{F}(j \times \omega) = [\mathbf{O}(j \times \omega) + \Delta\mathbf{O}(j \times \omega)]^{-1}$$
$$\mathbf{O}(j \times \omega) = -\omega^2 \times \mathrm{Re}(\mathbf{M}_u + \mathbf{M}_w) + j \times \omega \times \mathrm{Re}(\Delta\mathbf{C}_w + \Delta\mathbf{C}_u) + \mathrm{Re}((\mathbf{K}_\varepsilon + \mathbf{K}_\tau) + (\Delta\mathbf{K}_w + \Delta\mathbf{K}_u)) \tag{38}$$
$$\Delta\mathbf{O}(j \times \omega) = -\omega^2 \times j \times \mathrm{Im}(\mathbf{M}_u + \mathbf{M}_w) - \omega \times \mathrm{Im}(\Delta\mathbf{C}_w + \Delta\mathbf{C}_u) + j \times \mathrm{Im}((\mathbf{K}_\varepsilon + \mathbf{K}_\gamma) + (\Delta\mathbf{K}_w + \Delta\mathbf{K}_u))$$

And from Equation (38):

$$(\mathbf{O} + \Delta\mathbf{O}) \times \mathbf{F} = \mathbf{I}. \tag{39}$$

Pre-multiplying both sides of Equation (39) by $\mathbf{F}_0 = \mathbf{O}^{-1}$:

$$\mathbf{F}_0 \times (\mathbf{O} + \Delta\mathbf{O}) \times \mathbf{F} = \mathbf{F}_0. \tag{40}$$

With further manipulation of Equation (40):

$$\mathbf{F} = (\mathbf{I} + \mathbf{F}_0 \times \Delta\mathbf{O})^{-1} \times \mathbf{F}_0. \tag{41}$$

In Equation (41) $\mathbf{F}_0$ can be expressed in closed form by means of modal decomposition [45]. The beam frequency response function is:

$$\widetilde{F}(\chi_w,\chi_q,j\omega) = \begin{bmatrix} \mathbf{0} & \mathbf{\Psi}_w(\chi_w) & \mathbf{0} \end{bmatrix} \times \mathbf{F}(j\cdot\omega) \times \begin{bmatrix} \mathbf{0} & \mathbf{\Psi}_w(\chi_q) & \mathbf{0} \end{bmatrix}^T \tag{42}$$

Evaluation of $\widetilde{F}(\chi_w,\chi_q,j\omega)$ makes it possible to virtually estimate the damping behaviour of the beam under study, i.e., by using the previously defined function $z(j\omega)$ or by defining a new damping estimator $r(\chi_w,\chi_q,j\omega) = \left|\mathrm{Im}\left(\widetilde{F}(\chi_w,\chi_q,j\omega)\right)\right| / \left|\widetilde{F}(\chi_w,\chi_q,j\omega)\right|$, where $r \in \Re$, $r \in [0,1]$.

*5.4. Model Application Examples*

Two different beam architectures are presented as examples, B1 and B2. Their data are reported in Table 3 where $h_i$ is the thickness of the *i*-th beam layer. For both the examples $L = 1.1$ m and $g = 0.08$ m. The number of layers are $N = 3$ (B1) and $N = 7$ (B2). The constant parameters $K_u'$, $K_w'$, $C_u'$, $C_w'$ take into account the beam distributed viscoelastic boundary conditions (clamped-free) and are applied at $0 \leq \chi \leq 0.09$. Constant parameters $C_u''$ and $C_w''$ are used to define uniformly distributed viscous actions ($0 \leq \chi \leq 1$) that model the system inherent damping.

**Table 3.** Beam model data.

| Beam | $h_i$ (mm) | $E_i$ (GPa) | $G_i$ (GPa) | $\rho_i$ ($10^3$ kg/m$^3$) |
|---|---|---|---|---|
| | | **Beam Layer Data** | | |
| B1 | {25, 50, 25} | {9, 210, 9} | {3.5, 80, 3.5} | {1.9, 7.85, 1.9} |
| B2 | {10, 5, 10, 50, 10, 5, 10} | {9, 210, 9, 210, 9, 210, 9} | {3.5, 80, 3.5, 80, 3.5, 80, 3.5} | {1.9, 7.85, 1.9, 7.85, 1.9, 7.85, 1.9} |

| Beam | $K_w'$ ($10^{15}$ N/m$^4$) | $K_u'$ ($10^{15}$ N/m$^4$) | $C_w'$ ($10^5$ N·s/m$^4$) | $C_u'$ ($10^5$ N·s/m$^4$) | $C_w''$ ($10^5$ N·s/m$^4$) | $C_u''$ ($10^5$ N·s/m$^4$) |
|---|---|---|---|---|---|---|
| | | | **Constrain Parameters** | | | |
| B1 | 1 | 10 | 8 | 2 | 1.2 | 0.01 |
| B2 | 1 | 10 | 8 | 2 | 1.4 | 0.2 |

| Beam | $\varphi_i$ ($10^8$ N/m$^3$) | $\eta_i$ ($10^{10}$ N/m$^3$) |
|---|---|---|
| | **Interface Parameters** | |
| B1 | {6, 6} | {3, 3} |
| B2 | {8, 8, 8, 8, 8,8} | {5, 5, 5, 5, 5,5} |

The effect of introducing hysteretic dissipative actions at the layer interfaces can be observed by comparing the results reported in Figures 12 and 13. The real, the imaginary part of the inertance $\left(In(\chi_w,\chi_q,j\omega) = -\omega^2 \times \widetilde{F}(\chi_w,\chi_q,j\omega)\right)$ frequency response function (FRF), calculated at $\chi_w = 1$ and $\chi_q = 1$, and the damping estimator $r(jw)$ are plotted.

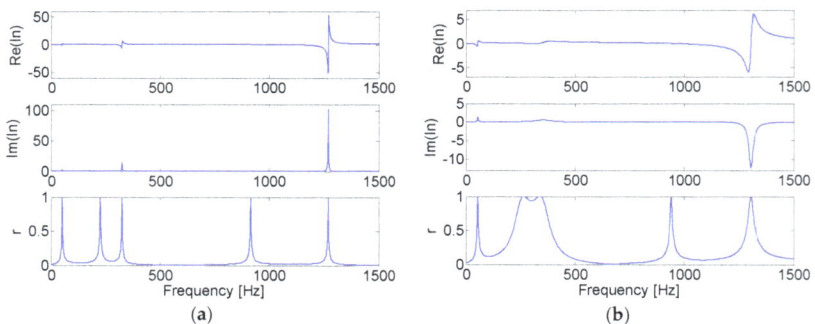

**Figure 12.** Inertance ($In(1,1,j\omega)$) real and imaginary part, and $r(jw)$ for beam B1$_0$ (a) and B1 (b).

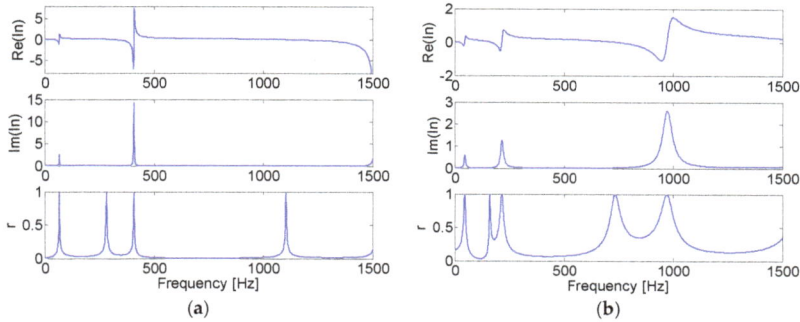

**Figure 13.** Inertance ($In(1,1,j\omega)$) real and imaginary part, and $r(j\omega)$ for beam B2$_0$ (**a**) and B2 (**b**).

Figures 12a and 13a show the results for beams B1$_0$ and B2$_0$ respectively, when no slipping occurs ($\varphi_i = 10^{16}$ N/m$^3$, $\eta_i = 0$, $\forall i$), Figures 12b and 13b show the results for beams B1 and B2 respectively, taking into account of interface slipping and hysteretic dissipation. Table 4 reports B1, B1$_0$, B2, B2$_0$ natural frequencies and damping ratios. B1 and B2 damping ratios are estimated by means of the Single Degree of Freedom (SDOF) circle fit method [45], while B1$_0$ and B2$_0$ damping ratios are normally obtained from within the solutions of a generalized eigenvalue problem [45].

**Table 4.** Natural frequencies and damping ratios.

| Beam | Natural Frequency (Hz) | | | Damping Ratio (%) | | |
|------|------|------|---------|------|------|--------|
| | 1 | 2 | 3 | 1 | 2 | 3 |
| B1$_0$ | 52.41 | 326.74 | 1269.16 | 2.32 | 0.43 | 0.109 |
| B1 | 54.04 | 352.48 | 1303.83 | 4.48 | 10.3 | 0.931 |
| B2$_0$ | 65.84 | 406.45 | 1291.1 | 1.95 | 0.36 | 0.0178 |
| B2 | 44.64 | 214.77 | 971.34 | 8.88 | 3.6 | 2.77 |

## 6. Conclusions

Eight different innovative composite solutions were experimentally investigated, and the results were compared; these findings were never before published. A significant increase of the damping behaviour is observed for all of these solutions with respect to the uncoated components, and also with respect to already known solutions, previously investigated by these authors and other researchers. The coating solution employing Al$_2$O$_3$ powder + matrix, made by screen printing and curing technology on Al alloy substrate, proved to be the most effective technology with respect to the aim of this work.

It can be outlined that the interface between substrate and coating heavily affects the effectiveness of the composite solution. Moreover, test results and comparison also outlined the influence of the substrate surface texture interface on the damped response. These results confirm the role of interface frictional actions in determining the composite damping behaviour. It should be outlined that the experimental estimate of the adhesion strength between coating and substrate can be evaluated by means of an adhesion test apparatus, and will be performed in future work.

An extended multi layered beam model was developed in order to design and optimize new, more effective coating solutions, engineered to maximize the damping contribution of frictional actions at the coating layer interfaces, or at the coating substrate interface. This model is based on high order multi-layer beam theories and takes into account of the contribution of interface frictional actions. The aim is to use the model as an experimental tool to identify the model parameters of the local dissipative actions acting at some interfaces between two different layers and as a design tool in order to explore new optimized coating solutions according to fixed engineering specifications. Some application examples of the model were presented. Nevertheless, experimental-numerical tools able to identify the complex elasto-hysteretic interface impedance must be developed in order to make this model suitable for engineering applications and will be presented in a future paper. Moreover, the application of this coating technology in real engineering applications, such as thin-walled components used in the automotive or automatic machine industry, should also be taken into account. Residual stresses at the interface between the coating layers and substrate may be taken into account in the multilayer beam model proposed in this work, since this contribution can be high and may be expected to play a major role in the adhesion behaviour of coated solutions when dealing with PVD based coating deposition technologies. It also appears that such extended beam models could also be used to experimentally identify this unknown residual stress field from measuring the deformed profile of a cantilever beam resulting from the application of a coated layer. Some analytical, numerical and experimental work is currently under development by our research team.

**Acknowledgments:** This study was developed within the CIRI-MAM with the contribution of the Regione Emilia Romagna, progetto POR Fesr Tecnopoli. Support from Andrea Zucchini and Marzocchi Pompe S.p.A., Casalecchio di Reno, Italy, is also kindly acknowledged.

**Author Contributions:** Stefano Amadori and Giuseppe Catania conceived, designed and performed the experiments; defined and implemented the numerical model, analyzed and critically discussed the numerical and experimental results and finally wrote the paper; Angelo Casagrande found the coating material composition, the manufacturing technology and performed the material experimental structural investigation.

**Conflicts of Interest:** The authors declare no conflict of interest.

## References

1. Lopez de Lacalle, L.N.; Gutierrez, A.; Lamikiz, A.; Fernandes, M.H.; Sanchez, J.A. Turning of Thick Thermal Spray Coatings. *J. Therm. Spray Technol.* **2001**, *10*, 249–254. [CrossRef]
2. Fernández-Abia, A.I.; Barreiro, J.; Lopez de Lacalle, L.N.; González-Madruga, D. Effect of mechanical pre-treatments in the behavior of nanostructured PVD-coated tools in turning. *Int. J. Adv. Manuf. Technol.* **2014**, *73*, 1119–1132. [CrossRef]
3. Rodríguez-Barrero, S.; Fernández-Larrinoa, J.; Azkona, I.; López de Lacalle, L.N.; Polvorosa, R. Enhanced Performance of Nanostructured Coatings for Drilling by Droplet Elimination. *Mater. Manuf. Process.* **2016**, *31*, 593–602. [CrossRef]
4. Fernandez-Valdivielso, A.; Lopez de Lacalle, LN.; Urbikain, G.; Rodriguez, A. Detecting the key geometrical features and grades of carbide inserts for the turning of nickel-based alloys concerning surface integrity. *J. Mech. Eng. Sci.* **2016**, *230*, 3725–3742. [CrossRef]
5. Polvorosab, R.; Suáreza, A.; Lopez de Lacalle, L.N.; Cerrill, I.; Wretlandc, A.; Veigaa, F. Tool wear on nickel alloys with different coolant pressures: Comparison of Alloy 718 and Waspaloy. *J. Manuf. Process.* **2017**, *26*, 44–56. [CrossRef]
6. Elosegui, I.; Alonso, U.; Lopez de Lacalle, L.N. PVD coatings for thread tapping of austempered ductile iron. *Int. J. Adv. Manuf. Technol.* **2017**, *91*, 2663–2672. [CrossRef]
7. Ghidelli, M.; Sebastiani, M.; Collet, C.; Guillemet, R. Determination of the elasticmoduli and residual stresses of freestanding Au-TiW bilayer thin films by nanoindentation. *Mater. Des.* **2016**, *106*, 436–445. [CrossRef]
8. Nix, W.D. Mechanical properties of thin films. *Metall. Trans. A* **1989**, *20*, 2217–2245. [CrossRef]
9. Oliver, W.C.; Pharr, G.M. Measurement of hardness and elastic modulus by instrumented indentation: Advances in understanding and refinements to methodology. *J. Mater. Res.* **2003**, *19*, 3–20. [CrossRef]
10. Tassini, N.; Pastias, S.; Lambrinou, K. Ceramic coatings: A phenomenological modeling for damping behavior related to microstructural features. *Mater. Sci. Eng. A* **2006**, *442*, 509–513. [CrossRef]
11. Yu, L.; Ma, Y.; Zhou, C.; Xu, H. Damping efficiency of the coating structure. *Int. J. Solids Struct.* **2005**, *42*, 3045–3058. [CrossRef]
12. Ustinov, A.I.; Movchan, B.A. A study of damping ability of tin-and yttrium-coated flat specimens of Ti-6%Al-4%V titanium alloy. *Strength Mater.* **2001**, *33*, 339–343. [CrossRef]

13. Rongong, J.A.; Goruppa, A.A.; Buravalla, V.R.; Tomlinson, G.R.; Jones, F.R. Plasma deposition of constrained layer damping coating. *J. Mech. Eng. Sci.* **2004**, *218*, 669–680. [CrossRef]

14. Blackwell, C.; Palazzotto, A.; George, T.J.; Cross, C.J. The evaluation of the damping characteristics of hard coating on titanium. *Shock Vib.* **2007**, *14*, 37–51. [CrossRef]

15. Casadei, F.; Bertoldi, K.; Clarke, D.R. Vibration damping of thermal barrier coatings containing ductile metallic layers. *ASME J. Appl. Mech.* **2014**, *81*, 101001. [CrossRef]

16. Du, G.; Tan, Z.; Ba, D.; Liu, K.; Sun, W.; Han, Q. Damping properties of a novel porous Mg-Al alloy coating prepared by arc ion plating. *Surf. Coat. Technol.* **2014**, *238*, 139–142. [CrossRef]

17. Ustinov, A.I.; Skorodzievskii, V.S. A study of the dissipative properties of homogeneous materials deposited as coatings part 2. Copper condensates with different microstructural characteristics. *Strength Mater.* **2008**, *40*, 275–277. [CrossRef]

18. Colorado, H.A.; Velez, J.; Salva, H.R.; Ghilarducci, A.A. Damping behavior of physical vapor-deposited TiN coatings on AISI 304 stainless steel and adhesion determination. *Mater. Sci. Eng. A* **2006**, *442*, 514–518. [CrossRef]

19. Khor, K.A.; Chia, C.T.; Gu, Y.W.; Boey, F.Y.C. High temperature damping behavior of plasma sprayed NiCoCrAlY coatings. *J. Therm. Spray Technol.* **2002**, *11*, 359–364. [CrossRef]

20. Wang, X.; Pei, Y.; Ma, Y. The effect of microstructure at interface between coating and substrate on damping capacity of coating systems. *Appl. Surf. Sci.* **2013**, *282*, 60–66. [CrossRef]

21. Kiretseu, M.; Hui, D.; Tomlinson, G. Advanced shock-resistant and vibration damping of nanoparticle-reinforced composite material. *Compos. Part B* **2008**, *39*, 128–138. [CrossRef]

22. Amadori, S.; Catania, G. Damping contributions of coatings to the viscoelastic behaviour of mechanical components. In Proceedings of the Surveillance 9 International Conference, Fez, Morocco, 22–24 May 2017; pp. 1–13.

23. Amadori, S.; Catania, G. Experimental evaluation of the damping properties and optimal modeling of coatings made by plasma-deposition techniques. In Proceedings of the 7th International Conference on Mechanics and Materials in Design Albufeira, Algarve, Portugal, 11–15 June 2017; INEGI/FEUP: Porto, Portugal, 2017; pp. 313–324.

24. Buravalla, V.R.; Remillat, C.; Rongrong, J.A.; Tomlinson, G.A. Advances in damping materials and technology. *Smart Mater. Bull.* **2001**, *8*, 10–13. [CrossRef]

25. Reed, S.A.; Palazzotto, A.N.; Baker, W.P. An experimental technique for the evaluation of strain dependent material properties of hard coatings. *Shock Vib.* **2008**, *15*, 697712. [CrossRef]

26. Pastias, S.; Saxton, C.; Shipton, M. Hard damping coatings: An experimental procedure for extraction of damping characteristics and modulus of elasticity. *Mater. Sci. Eng. A* **2004**, *370*, 412–416. [CrossRef]

27. Torvik, P.J. Determination of mechanical properties of non-linear coatings from measurements with coated beams. *Int. J. Solids Struct.* **2009**, *46*, 1066–1077. [CrossRef]

28. Averill, R.C.; Yip, Y.C. Development of simple robust finite elements based on refined theories for thick laminated beams. *Comput. Struct.* **1996**, *59*, 529546. [CrossRef]

29. Di Sciuva, M.; Gherlone, M.; Librescu, L. Implications of damaged interfaces and of other non-classical effects on the load carrying capacity of multilayered composite shallow shells. *Int. J. Non-Linear Mech.* **2002**, *37*, 851–867. [CrossRef]

30. Iurlaro, L.; Gherlone, M.; di Sciuva, M.; Tessler, A. Refined Zigzag Theory for laminated composite and sandwich plated derived from Reissner's Mixed Variational Theorem. *Compos. Struct.* **2015**, *133*, 809–817. [CrossRef]

31. Liu, D.; Li, X. An overall view of laminate theories based on the displacement hypothesis. *J. Compos. Mater.* **1996**, *30*, 1539–1561. [CrossRef]

32. Yong, S.L.; Feng, D.W.; Lukey, G.C.; van Deventer, J.S.J. Chemical characterization of the steel-geopolymeric gel interface. *Colloids Surf. A Physicochem. Eng. Asp.* **2007**, *302*, 411–423. [CrossRef]

33. Prud'Homme, E.; Michaud, P.; Joussein, E.; Clacens, J.; Arii-Clacens, S.; Sobrados, I.; Peyratout, C.; Smith, A.; Sanz, J.; Rossignol, S. Structural characterization of geomaterial foams—Thermal behavior. *J. Non-Cryst. Solids* **2011**, *357*, 3637–3647. [CrossRef]

34. Temuujin, J.; Minjihmaa, A.; Rickard, W.; Lee, M.; Williams, I.; van Riessen, A. Preparation of metacaolin based geopolymer coatings on metal substrates as thermal barriers. *Appl. Clay Sci.* **2009**, *46*, 265–270. [CrossRef]

35. He, J.; Zhang, J.; Yu, Y.; Zhang, G. The strength and microstructure of two geopolymers derived from metakaolin and red mud-fly ash admixture: A comparative study. *Constr. Build. Mater.* **2012**, *30*, 80–91. [CrossRef]

36. Rovnaník, P. Effect of curing temperature on the development of hard structure of metakaolin-based geopolymer. *Constr. Build. Mater.* **2010**, *24*, 1176–1183. [CrossRef]

37. Amadori, S.; Catania, G. Robust identification of the mechanical properties of viscoelatic non-standard materials by means of time and frequency domain experimental measurements. *Compos. Struct.* **2017**, *169*, 79–89. [CrossRef]

38. Nowick, A.S.; Berry, B.S. *Anelastic Relaxation in Crystalline Solids*; Academic Press Inc.: Cambridge, MA, USA, 1972.

39. Di Sciuva, M. Multilayered anisotropic plate models with continuous interlaminar stresses. *Compos. Struct.* **1992**, *22*, 149–167. [CrossRef]

40. Librescu, L.; Schmidt, R. A general linear theory of laminated composite shells featuring interlaminar bonding imperfections. *Int. J. Solid Struct.* **2001**, *38*, 3355–3375. [CrossRef]

41. Reddy, J.N. On refined theories of composite laminates. *Meccanica* **1990**, *25*, 230–238. [CrossRef]

42. Sun, C.T.; Whitney, J.M. Theory for the Dynamic Response of Laminated Plates. *AIAA J.* **1973**, *11*, 178–183. [CrossRef]

43. Sun, W.; Liu, Y.; Du, G. Analytical Modeling of Hard-Coating Cantilever Composite Plate considering the Material Nonlinearity of Hard Coating. *Math. Probl. Eng.* **2015**, *2015*, 978392. [CrossRef]

44. Wang, G.; Unal, A.; Zuo, Q.H. Modelling and Analysis of Multilayered Elastic Beam Using Spectral Finite Element Method. *J. Vib. Acoust.* **2016**, *138*, 041013. [CrossRef]

45. Ewins, D.J. *Modal Testing: Theory, Practice and Applications*, 2nd ed.; Research Studies Press, University of Michigan: Ann Arbor, MI, USA, 2000.

*coatings*

MDPI

*Article*

# Damping Oriented Design of Thin-Walled Mechanical Components by Means of Multi-Layer Coating Technology

**Giuseppe Catania and Matteo Strozzi \***

Department of Industrial Engineering, University of Bologna, Viale del Risorgimento 2, 40136 Bologna, Italy; giuseppe.catania@unibo.it
\* Correspondence: matteo.strozzi2@unibo.it; Tel.: +39-051-209-3909

Received: 29 December 2017; Accepted: 9 February 2018; Published: 13 February 2018

**Abstract:** The damping behaviour of multi-layer composite mechanical components, shown by recent research and application papers, is analyzed. A local dissipation mechanism, acting at the interface between any two different layers of the composite component, is taken into account, and a beam model, to be used for validating the known experimental results, is proposed. Multi-layer prismatic beams, consisting of a metal substrate and of some thin coated layers exhibiting variable stiffness and adherence properties, are considered in order to make it possible to study and validate this assumption. A dynamical model, based on a simple beam geometry but taking into account the previously introduced local dissipation mechanism and distributed visco-elastic constraints, is proposed. Some different application examples of specific multi-layer beams are considered, and some numerical examples concerning the beam free and forced response are described. The influence of the multilayer system parameters on the damping behaviour of the free and forced response of the composite beam is investigated by means of the definition of some damping estimators. Some effective multi-coating configurations, giving a relevant increase of the damping estimators of the coated structure with respect to the same uncoated structure, are obtained from the model simulation, and the results are critically discussed.

**Keywords:** damping; multi-layer beam; FGM; locally distributed viscosity

## 1. Introduction

Multi-coated thin-walled composite mechanical components, such as beams, plates, and shells, can be considered as an application of the Functionally Graded Material (FGM) technology, making it possible to obtain specific mechanical properties, such as high strength and stiffness, light inertia, and high damping. In most modern industrial applications, damping behaviour can be critical and may greatly affect design activities. In the aerospace field, high stiffness and high strength slender shell components, such as turbine blades, are required to show a limited free vibrational behaviour in standard operating conditions in order to increase the system life, to reduce the generated noise, and to maximize the machine efficiency. In this specific field, recent applications were explored, mainly based on experimental approaches. The influence of thin ceramic, polymeric, and metallic coatings, in some cases reinforced by carbon nanotubes, deposed on thin-walled components, on the damping behaviour of the obtained multi-layer composite systems was experimentally studied by some researchers [1–5]. In all of these works, it was found that the deposition of thin coatings can improve the damping behaviour of the composite system. Moreover, it was experimentally found in [6–10] that the dissipative actions in multi-layer architectures can be assumed as localized at the interfaces of the layers.

A global modelling approach of FGM composite components is given in [11], in which an experimental identification procedure of the mechanical properties of nonstandard composite materials by means of frequency domain measurements is reported. However, the global constitutive model of a specific composite material experimentally obtained cannot be used to obtain the model related to a different composite material solution, so that virtual prototyping of new solutions is de facto not allowed in principle. On the other hand, several works on multi-layer composite structure modelling can be found in the literature, but no attempt was made to investigate the dissipative mechanism acting at the interface between any two layers. In [12–17], multi-layer beam models, assuming polynomial first and higher order longitudinal displacement component, were proposed and validated by numerical comparisons with exact analytical solutions. Nevertheless, dissipation effects were not taken into account. A dynamical model that was able to deal with the dissipative actions influencing the damping properties of the system response and with the specific multilayer coating architecture was not proposed, to the authors' knowledge.

The damping behaviour of multi-layer composite beams is taken into account in this work and introduced in a beam model. The local dissipations, acting at the interface between any two layers of the composite beam, are investigated. The dissipation mechanism in a multi-layer structure is described by means of distributed viscous linear shear actions acting at the interface between two layers. The shear-strain local constitutive behaviour is described by defining continuous dissipation functions depending on the thickness and the viscosity at the interface. These parameters can be assumed to model the layer interface coupling actions, which mainly originated by chemical or mechanical coupling phenomena, and are associated with the different technologies employed to depose the layer coatings [18].

Although distributed viscous modelling of the internal shear dissipative actions was already published in known scientific literature, it must be taken into account that such an assumption can lead to misleading results, since while polymer-based materials may follow this behaviour, most metal and ceramic materials do not exhibit internal dissipative actions depending on strain velocity. Moreover, since a proportional damping model follows from this assumption, the theoretical modal damping ratio of a homogeneous component made of a viscoelastic material following the Kelvin-Voigt model linearly increases with respect to frequency, and this result is not supported by experimental findings [11,19]. Since it can be experimentally found that the free vibration modal damping ratio $\xi_k$ tends to slowly vary with respect to the mode natural frequency $\omega_k$ and order $k$ for homogeneous, uniform beam specimens made of viscoelastic material, high order material viscoelastic constitutive relationships [11] or fractional derivative order model-based viscoelastic relationships [20–23] were proposed in the past to overcome the limits of a simple Kelvin viscoelastic model.

In the approach proposed in this paper, viscous shear actions are mainly localized at the interface between any two different layers; they are defined by a $C^1$ function only depending on two parameters, and they are not distributed on the whole layer domain, making it possible to properly model and experimentally validate the modal damping ratio with respect to the modal natural frequency value. Multi-layer prismatic beams with distributed visco-elastic constraints that are subjected to distributed, dynamic load are considered, and a discrete dynamical beam model is introduced. The contribution of aerodynamic drag dissipative actions can be taken into account by means of modelling viscous constraint actions distributed along the whole beam length.

Some examples, consisting of multi-layer composite beams composed by a thick metal-based substrate with high strength and stiffness, and thin coatings, are considered. By varying the coating parameters, such as the number, thickness, and material properties of the layers, the interlaminar dissipation layer thickness, and the viscosity, the effect on the system damping estimate in the frequency range under interest is analysed by means of numerical simulations from within the proposed model. The sensitivity of the layer parameters on the damping behaviour of the structure is outlined. Some effective multilayer configurations associated with a relevant increase of the damping ratio with respect to the single layer solution are illustrated, and a critical discussion follows.

## 2. Multi-Layer Beam Modelling

In Figure 1, a scheme of a uniform, rectangular section, multi-layer composite beam is shown. Geometrical parameters are the beam length $L$, depth $b$, and thickness $h$; $N$ is the number of layers, $h_k$ is the $k$-th layer thickness, $x$ is the longitudinal, and $z$ is the transversal coordinate; $z_k$ is the $k$-th layer coordinate with respect to the bottom surface, $\xi$ and $\zeta$ are the dimensionless coordinates, defined by the relationships:

$$\xi = \frac{x}{L}, \quad \zeta = \frac{z}{h}, \quad \zeta_k = \frac{z_k}{h}, \quad k = 0, \dots, N \tag{1}$$

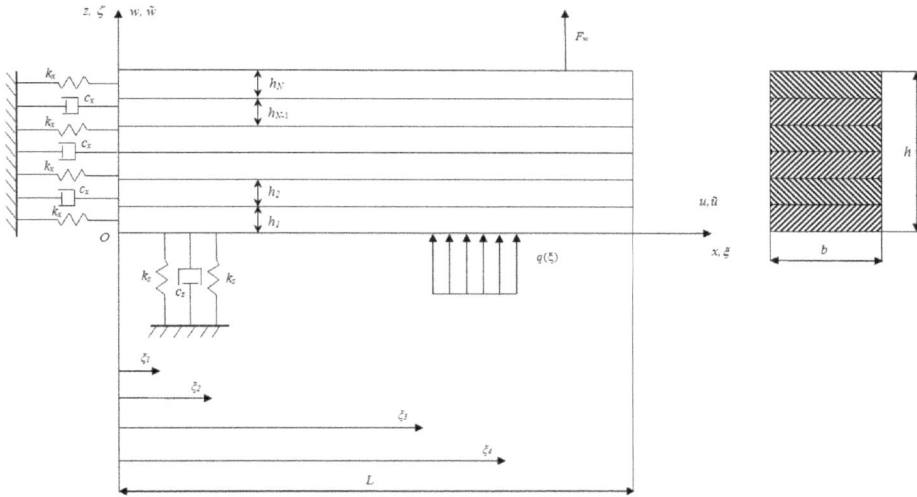

**Figure 1.** Multi-layer composite beam.

Kinematical parameters are the axial $\tilde{u}$ and transversal beam $\tilde{w}$ displacement components, $k_{x,z}$ is the stiffness of the longitudinal and transversal distributed elastic constraints, $c_{x,z}$ is the viscosity of the longitudinal and transversal distributed viscous constraints, $q$ and $F_w$ are the distributed and concentrated external transversal loads, $\rho_k$, $E_k$, and $G_k$ are the material, $k$-th layer, mass density, axial and shear moduli, and $t$ is the time coordinate.

The displacement field is defined in dimensionless form, i.e., $w = \tilde{w}/L$, $u = \tilde{u}/h$; by assuming $w$ to be independent of coordinate $\zeta$, this assumption is mainly valid in the low to medium excitation frequency range [24] taken into account in this work, and by assuming that $u$ varies with respect to $\zeta$ following a cubic polynomial function, whose linear parameter components vary in each layer in the form:

$$\begin{aligned} w(\xi, \zeta, t) &= w(\xi, t), \quad \frac{\partial w}{\partial \zeta} = 0 \\ u(\xi, \zeta, t) &= \alpha(\xi, t) + \beta(\xi, t) \cdot \zeta + \chi(\xi, t) \cdot \zeta^2 + \delta(\xi, t) \cdot \zeta^3 + \hat{a}_k(\xi, t) + \hat{b}_k(\xi, t) \cdot \zeta, \\ &\quad \zeta_{k-1} \le \zeta \le \zeta_k \end{aligned} \tag{2}$$

the following $(\alpha, \beta, \chi, \delta, \hat{a}_k, \hat{b}_k, w)$ unknown, $(2 \cdot N + 3)$ state variables result, where $\zeta_0 = \hat{a}_1 = \hat{b}_1 = \dot{\hat{b}}_1 = 0$.

Starting from Equation (2), in the hypothesis of small deformations, the axial normal strain for the $k$-th layer, $k = 1 \dots N$, is:

$$\varepsilon(\zeta) = \frac{h}{L} \cdot \left( \alpha' + \beta' \cdot \zeta + \chi' \cdot \zeta^2 + \delta' \cdot \zeta^3 + \hat{a}'_k + \hat{b}'_k \cdot \zeta \right), \zeta_{k-1} \le \zeta \le \zeta_k \tag{3}$$

with:

$$\frac{\partial(\ )}{\partial \zeta} = (\ )', \frac{\partial(\ )}{\partial t} = \dot{(\ )} \tag{4}$$

and the transverse shear strain for the $k$-th layer is:

$$\gamma(\zeta) = \beta + 2 \cdot \chi \cdot \zeta + 3 \cdot \delta \cdot \zeta^2 + \hat{b}_k + w', \quad \zeta_{k-1} \leq \zeta \leq \zeta_k \tag{5}$$

The transverse normal stress is neglected ($\sigma_{zz} = 0$), assuming the plane stress hypothesis.
The constitutive equation for the $k$-th layer of the beam, in the case of isotropic material, is:

$$\left\{ \begin{array}{c} \sigma_{xx} \\ \tau_{xz} \end{array} \right\} = \left\{ \begin{array}{c} \sigma \\ \tau \end{array} \right\} = \left[ \begin{array}{cc} E_k & 0 \\ 0 & G_k \end{array} \right] \left\{ \begin{array}{c} \varepsilon \\ \gamma \end{array} \right\} + \left[ \begin{array}{cc} 0 & 0 \\ 0 & G_k \cdot \eta_k(\zeta) \end{array} \right] \left\{ \begin{array}{c} \dot{\varepsilon} \\ \dot{\gamma} \end{array} \right\} \Rightarrow \left\{ \begin{array}{c} \sigma = E_k \cdot \varepsilon \\ \tau = G_k \cdot (\gamma + \eta_k(\zeta) \cdot \dot{\gamma}) \end{array} \right. , \quad \zeta_{k-1} \leq \zeta \leq \zeta_k \tag{6}$$

From Equations (3), (5) and (6):

$$\left( \begin{array}{ll} \sigma(\zeta) & = E_k \cdot \frac{h}{L} \cdot \left( \alpha' + \beta' \cdot \zeta + \chi' \cdot \zeta^2 + \delta' \cdot \zeta^3 + \hat{a}'_k + \hat{b}'_k \cdot \zeta \right) \\ \tau(\zeta) & = G_k \cdot (\gamma + \eta_k(\zeta) \cdot \dot{\gamma}) = \tau_e + \tau_a \\ \tau_e & = G_k \cdot \gamma = G_k \cdot \left( \beta + 2 \cdot \chi \cdot \zeta + 3 \cdot \delta \cdot \zeta^2 + \hat{b}_k + w' \right) \\ \tau_a & = G_k \cdot \eta_k(\zeta) \cdot \dot{\gamma} = G_k \cdot \eta_k(\zeta) \cdot \left( \dot{\beta} + 2 \cdot \dot{\chi} \cdot \zeta + 3 \cdot \dot{\delta} \cdot \zeta^2 + \dot{\hat{b}}_k + \dot{w}' \right) \end{array} \right. , \quad \zeta_{k-1} \leq \zeta \leq \zeta_k \tag{7}$$

in which $\eta_k(\zeta) \in C^1$, plotted in Figure 2, is assumed to model the viscous behaviour localized at the $k$-th interface:

$$\eta(\zeta) = \left\{ \begin{array}{ll} \eta_{k,u}(\zeta), & \zeta_k - s_k/h \leq \zeta \leq \zeta_k \\ \eta_{k+1,l}(\zeta), & \zeta_k \leq \zeta \leq \zeta_k + s_k/h \end{array} \right. \tag{8}$$

in which $s_k$ is the interlaminar dissipation layer thickness at the $k$-th interface.

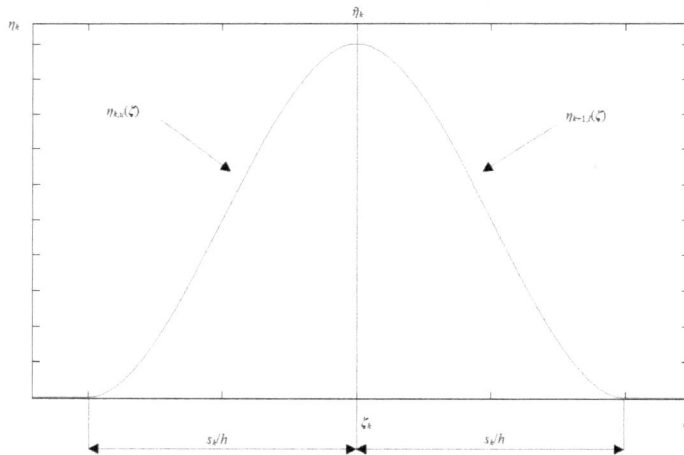

**Figure 2.** Dissipation function $\eta(\zeta)$ at the $k$-th interface.

The upper $\eta_{k,u}(\zeta)$ and lower $\eta_{k+1,l}(\zeta)$ contributions are given by means of interpolating polynomials:

$$\begin{array}{l} \eta_{k,u}(\zeta) = \tilde{\eta}_k \cdot \dfrac{2 \cdot \zeta^3 - 3 \cdot (2 \cdot \zeta_k - s_k/h) \cdot \zeta^2 + 6 \cdot \zeta_k \cdot (\zeta_k - s_k/h) \cdot \zeta - (\zeta_k - s_k/h)^2 \cdot (s_k/h + 2 \cdot \zeta_k)}{(\zeta_k - s_k/h)^3 - \zeta_k^3 + 3 \cdot \zeta_k \cdot (\zeta_k - s_k/h) \cdot (s_k/h)} \\[12pt] \eta_{k+1,l}(\zeta) = \tilde{\eta}_k \cdot \dfrac{2 \cdot \zeta^3 - 3 \cdot (2 \cdot \zeta_k + s_k/h) \cdot \zeta^2 + 6 \cdot \zeta_k \cdot (\zeta_k + s_k/h) \cdot \zeta + (\zeta_k + s_k/h)^2 \cdot (s_k/h - 2 \cdot \zeta_k)}{(\zeta_k + s_k/h)^3 - \zeta_k^3 - 3 \cdot \zeta_k \cdot (\zeta_k + s_k/h) \cdot (s_k/h)} \end{array} \tag{9}$$

satisfying the following conditions, $k = 1, \ldots, N - 1$:

$$
\frac{d\eta_{k,u}}{d\zeta}(\zeta_k) = \frac{d\eta_{k,u}}{d\zeta}(\zeta_k - s_k/h) = \frac{d\eta_{k+1,l}}{d\zeta}(\zeta_k) = \frac{d\eta_{k+1,l}}{d\zeta}(\zeta_k + s_k/h) = 0
$$
$$
\eta_{k+1,l}(\zeta_k + s_k/h) = \eta_{k,u}(\zeta_k - s_k/h) = 0, \quad \eta_k(\zeta_k) = \eta_{k,u}(\zeta_k) = \eta_{k+1,l}(\zeta_k) = \tilde{\eta}_k
$$

(10)

By imposing the continuity of the shear stress at the bottom surface ($\zeta = \zeta_0 = 0$):

$$
\tau(\zeta = 0) = G_1 \cdot \left( \beta + w' + \eta_1(0) \cdot \left( \dot{\beta} + \dot{w}' \right) \right) = G_1 \cdot \left( \beta + w' \right) = 0
$$

(11)

giving:

$$
\beta = -w'
$$

(12)

Substituting Equation (12) into Equation (7):

$$
\tau(\zeta) = G_k \cdot \left( 2 \cdot \chi \cdot \zeta + 3 \cdot \delta \cdot \zeta^2 + \hat{b}_k + \eta_k(\zeta) \cdot \left( 2 \cdot \dot{\chi} \cdot \zeta + 3 \cdot \dot{\delta} \cdot \zeta^2 + \dot{\hat{b}}_k \right) \right), \quad \zeta_{k-1} \le \zeta \le \zeta_k
$$

(13)

By imposing the continuity of the shear stress at the top surface ($\zeta = \zeta_N = 1$):

$$
\tau(\zeta = 1) = G_N \cdot \left( 2 \cdot \chi + 3 \cdot \delta + \hat{b}_N + \eta_N(1) \cdot \left( 2 \cdot \dot{\chi} + 3 \cdot \dot{\delta} + \dot{\hat{b}}_N \right) \right) = G_N \cdot \left( 2 \cdot \chi + 3 \cdot \delta + \hat{b}_N \right) = 0
$$

(14)

giving:

$$
\hat{b}_N = -2 \cdot \chi - 3 \cdot \delta
$$

(15)

By imposing the continuity of the shear stress at the $k$-th interface, $\tau(\zeta_k) = \tau\left( (\zeta_k)^- \right) = \tau\left( (\zeta_k)^+ \right)$, where $(\ )^- = \lim_{\Delta \to 0}(\ ) - \Delta$, $(\ )^+ = \lim_{\Delta \to 0}(\ ) + \Delta$:

$$
\tau(\zeta_k) = \tau\left( (\zeta_k)^- \right) = G_k \cdot \left( 2 \cdot \chi \cdot \zeta_k + 3 \cdot \delta \cdot \zeta_k^2 + \hat{b}_k + \tilde{\eta}_k \cdot \left( 2 \cdot \dot{\chi} \cdot \zeta_k + 3 \cdot \dot{\delta} \cdot \zeta_k^2 + \dot{\hat{b}}_k \right) \right) =
$$
$$
= G_{k+1} \cdot \left( 2 \cdot \chi \cdot \zeta_k + 3 \cdot \delta \cdot \zeta_k^2 + \hat{b}_{k+1} + \tilde{\eta}_k \cdot \left( 2 \cdot \dot{\chi} \cdot \zeta_k + 3 \cdot \dot{\delta} \cdot \zeta_k^2 + \dot{\hat{b}}_{k+1} \right) \right) = \tau\left( (\zeta_k)^+ \right)
$$

(16)

In the static case ($\dot{\chi} = \dot{\delta} = \dot{\hat{b}}_k = 0$), Equation (16) can be arranged as follows:

$$
\begin{cases}
G_2 \cdot \hat{b}_2 = (G_1 - G_2) \cdot (2 \cdot \chi \cdot \zeta_1 + 3 \cdot \delta \cdot \zeta_1^2) \\
G_3 \cdot \hat{b}_3 - G_2 \cdot \hat{b}_2 = (G_2 - G_3) \cdot (2 \cdot \chi \cdot \zeta_2 + 3 \cdot \delta \cdot \zeta_2^2) \\
\vdots \\
G_{k+1} \cdot \hat{b}_{k+1} - G_k \cdot \hat{b}_k = (G_k - G_{k+1}) \cdot (2 \cdot \chi \cdot \zeta_k + 3 \cdot \delta \cdot \zeta_k^2) \\
\vdots \\
G_N \cdot \hat{b}_N - G_{N-1} \cdot \hat{b}_{N-1} = (G_{N-1} - G_N) \cdot (2 \cdot \chi \cdot \zeta_{N-1} + 3 \cdot \delta \cdot \zeta_{N-1}^2)
\end{cases}
$$

(17)

By equating the sum of Equation (17) left side to the sum of the right side:

$$
\chi(\xi, \zeta) = \overline{\chi} \cdot \delta(\xi, \zeta), \quad \overline{\chi} = -\frac{3}{2} \cdot \frac{\left( \sum\limits_{k=1}^{N-1} \frac{G_k - G_{k+1}}{G_N} \cdot \zeta_k^2 + 1 \right)}{\left( \sum\limits_{k=1}^{N-1} \frac{G_k - G_{k+1}}{G_N} \cdot \zeta_k + 1 \right)}
$$

(18)

and the following iterative formula is obtained as well:

$$
\hat{b}_k = b_k \cdot \delta, \quad \begin{cases} b_1 = 0 \\ b_k = \frac{G_{k-1}}{G_k} \cdot b_{k-1} + \frac{G_{k-1} - G_k}{G_k} \cdot (2 \cdot \overline{\chi} + 3 \cdot \zeta_{k-1}) \cdot \zeta_{k-1}, \quad k = 2, \ldots, N \end{cases}
$$

(19)

It can be easily found that this result, collected in Equations (18) and (19), is also valid in the general dynamic case, since evaluating $\tau\left(\left(\zeta_k\right)^+\right)$, Equation (16) is still validated:

$$
\begin{aligned}
\tau(\zeta_k) &= \tau\left(\left(\zeta_k\right)^-\right) = G_k \cdot \left(2 \cdot \overline{\chi} \cdot \zeta_k + 3 \cdot \zeta_k^2 + b_k\right) \cdot \left(\delta + \widetilde{\eta}_k \cdot \dot{\delta}\right), \\
\tau(\zeta_k) &= \tau\left(\left(\zeta_k\right)^+\right) = G_{k+1} \cdot \left(2 \cdot \overline{\chi} \cdot \zeta_k + 3 \cdot \zeta_k^2 + b_{k+1}\right) \cdot \left(\delta + \widetilde{\eta}_k \cdot \dot{\delta}\right) \\
&= \left(\delta + \widetilde{\eta}_k \cdot \dot{\delta}\right) \cdot \left(G_{k+1} \cdot \left(2 \cdot \overline{\chi} \cdot \zeta_k + 3 \cdot \zeta_k^2\right) + G_k \cdot b_k + \left(G_k - G_{k+1}\right) \cdot \left(2 \cdot \overline{\chi} \cdot \zeta_k + 3 \cdot \zeta_k^2\right)\right) \\
&= G_k \cdot \left(\delta + \widetilde{\eta}_k \cdot \dot{\delta}\right) \cdot \left(2 \cdot \overline{\chi} \cdot \zeta_k + 3 \cdot \zeta_k^2 + b_k\right) = \tau\left(\left(\zeta_k\right)^-\right)
\end{aligned}
\tag{20}
$$

By imposing the continuity of the axial displacement at the $k$-th interface:

$$
b_k \cdot \zeta_k \cdot \delta + \hat{a}_k = b_{k+1} \cdot \zeta_k \cdot \delta + \hat{a}_{k+1}
\tag{21}
$$

the following iterative formula is obtained:

$$
\hat{a}_k = a_k \cdot \delta, \quad
\begin{cases}
a_1 = 0 \\
a_k = a_{k-1} + \left(b_{k-1} - b_k\right) \cdot \zeta_{k-1}
\end{cases}, \quad k = 2, \ldots, N.
\tag{22}
$$

From Equations (12), (15), (18), (19) and (22):

$$
u(\zeta) = \alpha - w' \cdot \zeta + \left(\zeta^3 + \overline{\chi} \cdot \zeta^2 + b_k \cdot \zeta + a_k\right) \cdot \delta, \quad \zeta_{k-1} \le \zeta \le \zeta_k
\tag{23}
$$

The strain components are:

$$
\begin{aligned}
\varepsilon(\zeta) &= \tfrac{h}{L} \cdot \left(\alpha' - \zeta \cdot w'' + \left(\zeta^3 + \overline{\chi} \cdot \zeta^2 + b_k \cdot \zeta + a_k\right) \cdot \delta'\right) \\
\gamma(\zeta) &= \left(3 \cdot \zeta^2 + 2 \cdot \overline{\chi} \cdot \zeta + b_k\right) \cdot \delta
\end{aligned}, \quad \zeta_{k-1} \le \zeta \le \zeta_k
\tag{24}
$$

and the stress components are:

$$
\begin{aligned}
\sigma(\zeta) &= E_k \cdot \tfrac{h}{L} \cdot \left(\alpha' - \zeta \cdot w'' + \left(\zeta^3 + \overline{\chi} \cdot \zeta^2 + b_k \cdot \zeta + a_k\right) \cdot \delta'\right) \\
\tau(\zeta) &= G_k \cdot \left(3 \cdot \zeta^2 + 2 \cdot \overline{\chi} \cdot \zeta + b_k\right) \cdot \left(\delta + \eta_k(\zeta) \cdot \dot{\delta}\right)
\end{aligned}, \quad \zeta_{k-1} \le \zeta \le \zeta_k.
\tag{25}
$$

Three independent model state scalar variables result and are collected: $\boldsymbol{\phi}(\xi,t) = [\alpha \; w \; \delta]^T$.
The axial displacement, and the normal and shear strains, can be expressed as a function of $\boldsymbol{\phi}(\xi,t)$:

$$
\begin{aligned}
u(\zeta) &= \mathbf{A}_k \cdot \mathbf{L}_1(\boldsymbol{\phi}), \quad \zeta_{k-1} \le \zeta \le \zeta_k \\
&\mathbf{A}_k(\zeta) = \begin{bmatrix} 1 & -\zeta & \left(\zeta^3 + \overline{\chi} \cdot \zeta^2 + b_k \cdot \zeta + a_k\right) \end{bmatrix}, \quad \mathbf{L}_1(\ ) = diag\left(\begin{bmatrix} (\ ) & (\ )' & (\ ) \end{bmatrix}^T\right) \\
\varepsilon(\zeta) &= \tfrac{h}{L} \cdot \mathbf{A}_k \cdot \mathbf{L}_2(\boldsymbol{\phi}), \quad \mathbf{L}_2(\ ) = (\mathbf{L}_1)' = diag\left(\begin{bmatrix} (\ )' & (\ )'' & (\ )' \end{bmatrix}^T\right) \\
\gamma(\zeta) &= \begin{bmatrix} 0 & 0 & B_k \end{bmatrix} \cdot \boldsymbol{\phi}, \quad B_k(\zeta) = 3 \cdot \zeta^2 + 2 \cdot \overline{\chi} \cdot \zeta + b_k
\end{aligned}
\tag{26}
$$

The equations of motion can be obtained by minimizing the total potential $\Pi$, given by the sum of the elastic deformation energy $U_{el}$, the interlaminar dissipation energy $U_{diss}$, the inertial force work $W_{in}$, the external force work $W_{ext}$, and the potential associated with the visco-elastic constraints $\Delta\Pi$:

$$
\Pi(\boldsymbol{\phi}) = U_{el} + U_{diss} + W_{in} + W_{ext} + \Delta\Pi \to \min \quad \Rightarrow \quad \frac{\partial \Pi}{\partial \boldsymbol{\phi}} = 0
\tag{27}
$$

$$
\begin{aligned}
U_{el} &= \tfrac{1}{2} \cdot b \cdot h \cdot \int_0^1 \sum_{k=1}^N \int_{\zeta_{k-1}}^{\zeta_k} \left(\tfrac{h^2}{L} \cdot E_k \cdot (\mathbf{L}_2 \cdot \boldsymbol{\phi})^T \cdot \hat{A}_k \cdot \mathbf{L}_2 \cdot \boldsymbol{\phi} + L \cdot G_k \cdot B_k^2 \cdot \boldsymbol{\phi}^T \cdot diag\left(\begin{bmatrix} 0 & 0 & 1 \end{bmatrix}^T\right) \cdot \boldsymbol{\phi}\right) \cdot d\zeta \cdot d\xi \\
&\hat{A}_k(\zeta) = A_k^T \cdot A_k = \begin{bmatrix} 1 & -\zeta & \left(\zeta^3 + \overline{\chi} \cdot \zeta^2 + b_k \cdot \zeta + a_k\right) \\ \cdots & \zeta^2 & \left(-\zeta^4 - \overline{\chi} \cdot \zeta^3 - b_k \cdot \zeta^2 - a_k \cdot \zeta\right) \\ sym & \cdots & \left(\zeta^3 + \overline{\chi} \cdot \zeta^2 + b_k \cdot \zeta + a_k\right)^2 \end{bmatrix}
\end{aligned}
\tag{28}
$$

$$U_{diss} = b \cdot h \cdot L \cdot \int\limits_0^1 \sum_{k=1}^N G_k \cdot \int\limits_{\zeta_{k-1}}^{\zeta_k} \eta_k \cdot \boldsymbol{\phi}^T \cdot diag\left(\begin{bmatrix} 0 & 0 & B_k^2 \end{bmatrix}^T\right) \cdot \dot{\boldsymbol{\phi}} \cdot d\zeta \cdot d\xi \tag{29}$$

$$W_{in} = b \cdot h \cdot L \cdot \int\limits_0^1 \sum_{k=1}^N \rho_k \cdot \int\limits_{\zeta_{k-1}}^{\zeta_k} \left( h^2 \cdot (L_1 \cdot \boldsymbol{\phi})^T \cdot \hat{A}_k \cdot L_1 \cdot \ddot{\boldsymbol{\phi}} + L^2 \cdot \boldsymbol{\phi}^T \cdot diag\left(\begin{bmatrix} 0 & 1 & 0 \end{bmatrix}^T\right) \cdot \ddot{\boldsymbol{\phi}} \right) \cdot d\zeta \cdot d\xi \tag{30}$$

$$W_{ext} = -L^2 \cdot \int\limits_{\xi_3}^{\xi_4} q \cdot \boldsymbol{\phi}^T(\xi) \cdot \begin{bmatrix} 0 & 1 & 0 \end{bmatrix}^T \cdot d\xi - L \cdot F_w \cdot \boldsymbol{\phi}^T(\overline{\xi}) \cdot \begin{bmatrix} 0 & 1 & 0 \end{bmatrix}^T \tag{31}$$

$$\Delta\Pi = \tfrac{1}{2} \cdot b \cdot L \cdot \int\limits_{\xi_1}^{\xi_2} \left( h^3 \cdot k_x \cdot (L_1 \cdot \boldsymbol{\phi})^T \cdot S \cdot L_1 \cdot \boldsymbol{\phi} + L^2 \cdot k_z \cdot \boldsymbol{\phi}^T \cdot diag\left(\begin{bmatrix} 0 & 1 & 0 \end{bmatrix}^T\right) \cdot \boldsymbol{\phi} \right) \cdot d\xi$$
$$+ b \cdot L \cdot \int\limits_{\xi_1}^{\xi_2} \left( h^3 \cdot c_x \cdot (L_1 \cdot \boldsymbol{\phi})^T \cdot S \cdot L_1 \cdot \dot{\boldsymbol{\phi}} + L^2 \cdot c_z \cdot \boldsymbol{\phi}^T \cdot diag\left(\begin{bmatrix} 0 & 1 & 0 \end{bmatrix}^T\right) \cdot \boldsymbol{\phi} \right) \cdot d\xi \,, \quad S = \sum_{k=1}^N \int\limits_{\zeta_{k-1}}^{\zeta_k} \hat{A}_k \cdot d\zeta \tag{32}$$

## 3. Numerical Discretization

Since the system Equation (27) is expressed by means of integro-differential equations that cannot be generally solved in closed form, then a model discretization is proposed, by means of the following hypothesis:

$$\boldsymbol{\phi}(\xi, t) = \begin{bmatrix} \alpha(\xi, t) \\ w(\xi, t) \\ \delta(\xi, t) \end{bmatrix} \approx \mathbf{N}(\xi) \cdot \mathbf{Y}(t) \tag{33}$$

in which $\mathbf{Y}(t)$ is unknown and $\mathbf{N}(\xi)$ is assumed to be known, and expressed by means of harmonic shape functions in the variable $\xi$ having argument $i \cdot \pi, \ i \in \mathbb{N}$:

$$\underset{3 \times n}{\mathbf{N}(\xi)} = \begin{bmatrix} \mathbf{N}_\alpha(\xi) & 0 & 0 \\ 0 & \mathbf{N}_w(\xi) & 0 \\ 0 & 0 & \mathbf{N}_\delta(\xi) \end{bmatrix}$$
$$\mathbf{N}_\alpha(\xi) = \sqrt{2} \cdot \begin{bmatrix} \tfrac{1}{\sqrt{2}} & \sin(\pi \cdot \xi) & \cos(\pi \cdot \xi) & \cdots & \sin(n_\alpha \cdot \pi \cdot \xi) & \cos(n_\alpha \cdot \pi \cdot \xi) \end{bmatrix} \tag{34}$$
$$\mathbf{N}_w(\xi) = \sqrt{2} \cdot \begin{bmatrix} \tfrac{1}{\sqrt{2}} & \sqrt{6} \cdot \left(\xi - \tfrac{1}{2}\right) & \sin(\pi \cdot \xi) & \cos(\pi \cdot \xi) & \cdots & \sin(n_w \cdot \pi \cdot \xi) & \cos(n_w \cdot \pi \cdot \xi) \end{bmatrix}$$
$$\mathbf{N}_\delta(\xi) = \sqrt{2} \cdot \begin{bmatrix} \sin(\pi \cdot \xi) & \cos(\pi \cdot \xi) & \cdots & \sin(n_\delta \cdot \pi \cdot \xi) & \cos(n_\delta \cdot \pi \cdot \xi) \end{bmatrix}$$

Equation (34) makes it possible to model three plane beam rigid body motions, so that the total number $n_{\text{tot}}$ of system discrete degrees of freedom is:

$$n_{\text{tot}} = 2 \cdot (n_\alpha + n_w + n_\delta) + 3 \tag{35}$$

From Equations (26), (32), and (33):

$$u(\zeta) = A_k \cdot \begin{bmatrix} N_\alpha & 0 & 0 \\ 0 & N'_w & 0 \\ 0 & 0 & N_\delta \end{bmatrix} \cdot \mathbf{Y}, \quad \varepsilon(\zeta) = \tfrac{h}{L} \cdot A_k \cdot \begin{bmatrix} N'_\alpha & 0 & 0 \\ 0 & N''_w & 0 \\ 0 & 0 & N'_\delta \end{bmatrix} \cdot \mathbf{Y}$$

$$\gamma(\zeta) = B_k \cdot \begin{bmatrix} 0 & 0 & N_\delta \end{bmatrix} \cdot \mathbf{Y}, \quad \sigma(\zeta) = E_k \cdot \tfrac{h}{L} \cdot A_k \cdot \begin{bmatrix} N'_\alpha & 0 & 0 \\ 0 & N''_w & 0 \\ 0 & 0 & N'_\delta \end{bmatrix} \cdot \mathbf{Y} \quad , \quad \zeta_{k-1} \le \zeta \le \zeta_k \tag{36}$$

$$\tau(\zeta) = G_k \cdot B_k \cdot \left( \begin{bmatrix} 0 & 0 & N_\delta \end{bmatrix} \cdot \mathbf{Y} + \eta(\zeta) \cdot \begin{bmatrix} 0 & 0 & N_\delta \end{bmatrix} \cdot \dot{\mathbf{Y}} \right)$$

The elastic deformation energy is:

$$U_{el} = \frac{1}{2} \cdot \mathbf{Y}^T \cdot (\mathbf{K_\sigma} + \mathbf{K_\tau}) \cdot \mathbf{Y}$$

$$\mathbf{K_\sigma} = \frac{b \cdot h^3}{L} \cdot \begin{bmatrix} S_{E_{11}} \cdot \int_0^1 {\mathbf{N}_\alpha'}^T \cdot \mathbf{N}_\alpha' \cdot d\xi & S_{E_{12}} \cdot \int_0^1 {\mathbf{N}_\alpha'}^T \cdot \mathbf{N}_w'' \cdot d\xi & S_{E_{13}} \cdot \int_0^1 {\mathbf{N}_\alpha'}^T \cdot \mathbf{N}_\delta' \cdot d\xi \\ S_{E_{21}} \cdot \int_0^1 {\mathbf{N}_w''}^T \cdot \mathbf{N}_\alpha' \cdot d\xi & S_{E_{22}} \cdot \int_0^1 {\mathbf{N}_w''}^T \cdot \mathbf{N}_w'' \cdot d\xi & S_{E_{23}} \cdot \int_0^1 {\mathbf{N}_w''}^T \cdot \mathbf{N}_\delta' \cdot d\xi \\ S_{E_{31}} \cdot \int_0^1 {\mathbf{N}_\delta'}^T \cdot \mathbf{N}_\alpha' \cdot d\xi & S_{E_{32}} \cdot \int_0^1 {\mathbf{N}_\delta'}^T \cdot \mathbf{N}_w'' \cdot d\xi & S_{E_{33}} \cdot \int_0^1 {\mathbf{N}_\delta'}^T \cdot \mathbf{N}_\delta' \cdot d\xi \end{bmatrix} \tag{37}$$

$$\mathbf{K_\tau} = b \cdot h \cdot L \cdot \sum_{k=1}^N G_k \cdot \int_{\zeta_{k-1}}^{\zeta_k} B_k^2 \cdot d\zeta \cdot \begin{bmatrix} 0 & 0 & 0 \\ 0 & 0 & 0 \\ 0 & 0 & \int_0^1 \mathbf{N}_\delta^T \cdot \mathbf{N}_\delta \cdot d\xi \end{bmatrix}, \quad \mathbf{S_E} = \sum_{k=1}^N E_k \cdot \int_{\zeta_{k-1}}^{\zeta_k} \hat{\mathbf{A}}_k \cdot d\zeta$$

The dissipation energy contribution is:

$$U_{diss} = \mathbf{Y}^T \cdot \mathbf{C} \cdot \dot{\mathbf{Y}}, \quad \mathbf{C} = b \cdot h \cdot L \cdot \sum_{k=1}^{N-1} G_k \cdot \int_{\zeta_k - s_k}^{\zeta_k + s_k} \eta(\zeta) \cdot B_k^2 \cdot d\zeta \cdot \int_0^1 \begin{bmatrix} 0 & 0 & 0 \\ 0 & 0 & 0 \\ 0 & 0 & \mathbf{N}_\delta^T \cdot \mathbf{N}_\delta \end{bmatrix} \cdot d\xi \tag{38}$$

The contribution of the inertial forces is:

$$W_{in} = \mathbf{Y}^T \cdot (\mathbf{M_u} + \mathbf{M_w}) \cdot \ddot{\mathbf{Y}}$$

$$\mathbf{M_u} = b \cdot h^3 \cdot L \cdot \begin{bmatrix} S_{\rho_{11}} \cdot \int_0^1 \mathbf{N}_\alpha^T \cdot \mathbf{N}_\alpha \cdot d\xi & S_{\rho_{12}} \cdot \int_0^1 \mathbf{N}_\alpha^T \cdot \mathbf{N}_w' \cdot d\xi & S_{\rho_{13}} \cdot \int_0^1 \mathbf{N}_\alpha^T \cdot \mathbf{N}_\delta \cdot d\xi \\ S_{\rho_{21}} \cdot \int_0^1 {\mathbf{N}_w'}^T \cdot \mathbf{N}_\alpha \cdot d\xi & S_{\rho_{22}} \cdot \int_0^1 {\mathbf{N}_w'}^T \cdot \mathbf{N}_w' \cdot d\xi & S_{\rho_{23}} \cdot \int_0^1 {\mathbf{N}_w'}^T \cdot \mathbf{N}_\delta \cdot d\xi \\ S_{\rho_{31}} \cdot \int_0^1 \mathbf{N}_\delta^T \cdot \mathbf{N}_\alpha \cdot d\xi & S_{\rho_{32}} \cdot \int_0^1 \mathbf{N}_\delta^T \cdot \mathbf{N}_w' \cdot d\xi & S_{\rho_{33}} \cdot \int_0^1 \mathbf{N}_\delta^T \cdot \mathbf{N}_\delta \cdot d\xi \end{bmatrix} \tag{39}$$

$$\mathbf{M_w} = b \cdot h \cdot L^3 \cdot \sum_{k=1}^N \rho_k \cdot (\zeta_k - \zeta_{k-1}) \cdot \begin{bmatrix} 0 & 0 & 0 \\ 0 & \int_0^1 \mathbf{N}_w^T \cdot \mathbf{N}_w \cdot d\xi & 0 \\ 0 & 0 & 0 \end{bmatrix}, \quad \mathbf{S_\rho} = \sum_{k=1}^N \rho_k \cdot \int_{\zeta_{k-1}}^{\zeta_k} \hat{\mathbf{A}}_k \cdot d\zeta$$

The contribution of the external forces is:

$$W_{ext} = -\mathbf{Y}^T \cdot L \cdot \left( F_w \cdot \begin{bmatrix} 0 \\ \mathbf{N}_w^T(\bar{\xi}) \\ 0 \end{bmatrix} + L \cdot \int_{\xi_3}^{\xi_4} \mathbf{N}_w^T(\xi) \cdot q \cdot d\xi \right) = -\mathbf{Y}^T \cdot \mathbf{F} \tag{40}$$

The potential energy associated with the visco-elastic constraints is:

$$\Delta\Pi = \frac{1}{2} \cdot \mathbf{Y}^T \cdot (\Delta\mathbf{K_u} + \Delta\mathbf{K_w}) \cdot \mathbf{Y} + \mathbf{Y}^T \cdot (\Delta\mathbf{C_u} + \Delta\mathbf{C_w}) \cdot \dot{\mathbf{Y}}$$

$$\Delta\mathbf{K_u} = b \cdot h^3 \cdot L \cdot \begin{bmatrix} S_{11} \int_{\xi_1}^{\xi_2} k_x \cdot \mathbf{N}_\alpha^T \cdot \mathbf{N}_\alpha \cdot d\xi & S_{12} \int_{\xi_1}^{\xi_2} k_x \cdot \mathbf{N}_\alpha^T \cdot \mathbf{N}_w' \cdot d\xi & S_{13} \int_{\xi_1}^{\xi_2} k_x \cdot \mathbf{N}_\alpha^T \cdot \mathbf{N}_\delta \cdot d\xi \\ S_{21} \int_{\xi_1}^{\xi_2} k_x \cdot {\mathbf{N}_w'}^T \cdot \mathbf{N}_\alpha \cdot d\xi & S_{22} \int_{\xi_1}^{\xi_2} k_x \cdot {\mathbf{N}_w'}^T \cdot \mathbf{N}_w' \cdot d\xi & S_{23} \int_{\xi_1}^{\xi_2} k_x \cdot {\mathbf{N}_w'}^T \cdot \mathbf{N}_\delta \cdot d\xi \\ S_{31} \int_{\xi_1}^{\xi_2} k_x \cdot \mathbf{N}_\delta^T \cdot \mathbf{N}_\alpha \cdot d\xi & S_{32} \int_{\xi_1}^{\xi_2} k_x \cdot \mathbf{N}_\delta^T \cdot \mathbf{N}_w' \cdot d\xi & S_{33} \int_{\xi_1}^{\xi_2} k_x \cdot \mathbf{N}_\delta^T \cdot \mathbf{N}_\delta \cdot d\xi \end{bmatrix}$$

$$\Delta\mathbf{K_w} = b \cdot L^3 \cdot \begin{bmatrix} 0 & 0 & 0 \\ 0 & \int_{\xi_1}^{\xi_2} k_z \cdot \mathbf{N}_w^T \cdot \mathbf{N}_w \cdot d\xi & 0 \\ 0 & 0 & 0 \end{bmatrix}, \quad \Delta\mathbf{C_w} = b \cdot L^3 \cdot \begin{bmatrix} 0 & 0 & 0 \\ 0 & \int_{\xi_1}^{\xi_2} c_z \cdot \mathbf{N}_w^T \cdot \mathbf{N}_w \cdot d\xi & 0 \\ 0 & 0 & 0 \end{bmatrix} \tag{41}$$

$$\Delta\mathbf{C_u} = b \cdot h^3 \cdot L \cdot \begin{bmatrix} S_{11} \int_{\xi_1}^{\xi_2} c_x \cdot \mathbf{N}_\alpha^T \cdot \mathbf{N}_\alpha \cdot d\xi & S_{12} \int_{\xi_1}^{\xi_2} c_x \cdot \mathbf{N}_\alpha^T \cdot \mathbf{N}_w' \cdot d\xi & S_{13} \int_{\xi_1}^{\xi_2} c_x \cdot \mathbf{N}_\alpha^T \cdot \mathbf{N}_\delta \cdot d\xi \\ S_{21} \int_{\xi_1}^{\xi_2} c_x \cdot {\mathbf{N}_w'}^T \cdot \mathbf{N}_\alpha \cdot d\xi & S_{22} \int_{\xi_1}^{\xi_2} c_x \cdot {\mathbf{N}_w'}^T \cdot \mathbf{N}_w' \cdot d\xi & S_{23} \int_{\xi_1}^{\xi_2} c_x \cdot {\mathbf{N}_w'}^T \cdot \mathbf{N}_\delta \cdot d\xi \\ S_{31} \int_{\xi_1}^{\xi_2} c_x \cdot \mathbf{N}_\delta^T \cdot \mathbf{N}_\alpha \cdot d\xi & S_{32} \int_{\xi_1}^{\xi_2} c_x \cdot \mathbf{N}_\delta^T \cdot \mathbf{N}_w' \cdot d\xi & S_{33} \int_{\xi_1}^{\xi_2} c_x \cdot \mathbf{N}_\delta^T \cdot \mathbf{N}_\delta \cdot d\xi \end{bmatrix}$$

From Equations (27) and (37)–(41):

$$(\mathbf{M_u} + \mathbf{M_w}) \cdot \ddot{\mathbf{Y}} + (\mathbf{C} + \Delta\mathbf{C_u} + \Delta\mathbf{C_w}) \cdot \dot{\mathbf{Y}} + (\mathbf{K_\sigma} + \mathbf{K_\tau} + \Delta\mathbf{K_u} + \Delta\mathbf{K_w}) \cdot \mathbf{Y} = \mathbf{F} \tag{42}$$

## 4. Damping Behaviour Estimate

From Equation (42), the following eigenproblem can be obtained:

$$\left( \lambda_r^2 \cdot (\mathbf{M_u} + \mathbf{M_w}) + \lambda_r \cdot (\mathbf{C} + \Delta\mathbf{C_u} + \Delta\mathbf{C_w}) + \mathbf{K_\sigma} + \mathbf{K_\tau} + \Delta\mathbf{K_u} + \Delta\mathbf{K_w} \right) \cdot \mathbf{\Delta}_r = 0 \tag{43}$$

in which $2 \cdot n_{tot}$ complex conjugate $\lambda_r$ eigenvalues and $\mathbf{\Delta}_r$ eigenvectors are expected to result.

From $\lambda_r$ $r$-th eigenvalue, natural circular frequency $f_r$ and damping ratio $\zeta_r$ can be evaluated as follows:

$$f_r = \frac{|\lambda_r|}{2\pi}, \quad \zeta_r = -\frac{\Re(\lambda_r)}{|\lambda_r|} \tag{44}$$

The $\zeta_r$ values may be considered as a useful system damping estimate in a local frequency range close to $r$-th natural frequency $f_r$.

A better damping estimate, depending on the input-output coordinate choice and on a frequency range $[f_{min}, f_{max}]$, may be obtained from within the evaluation of the system frequency response function $H(j \cdot 2\pi f)$ related to input in $x_{force}$ and output in $x_{response}$.

From Equation (40), by defining a discrete vector of equivalent force $\mathbf{F}$ related to unitary excitation $1 \cdot e^{j2\pi f}$ at $x_{force}$ coordinate:

$$\mathbf{F} = L \cdot \begin{bmatrix} \mathbf{0}^T & N_w(\zeta_{force}) & \mathbf{0}^T \end{bmatrix}^T \tag{45}$$

$Y_s$ $s$-th response component of vector $\mathbf{Y}$ may be easily evaluated by means of the modal approach [25]:

$$Y_s = \frac{\sum\limits_{r=1}^{2 \cdot ntot} \Delta_{s,r} \cdot \sum\limits_{i=1}^{ntot} \Delta_{i,r} \cdot F_i}{j \cdot 2\pi f - \lambda_r} \tag{46}$$

and the complex frequency response function related to $x_{force}$ and $x_{response}$ is:

$$H\left(j \cdot 2\pi f; x_{force}, x_{response}\right) = \begin{bmatrix} 0 & 1 & 0 \end{bmatrix} \cdot \mathbf{N}(\zeta_{resp}) \cdot \mathbf{Y} \tag{47}$$

Since the imaginary part of $H$ is mainly responsible of the damping behaviour, a normalized scalar function may be defined as:

$$d(j \cdot 2\pi f) = \frac{|\Im(H(j \cdot 2\pi f))|}{|H(j \cdot 2\pi f)|} \tag{48}$$

in which $d \in [0, 1]$ is defined for every frequency value and can be plotted in a limited frequency range related to a specific engineering field of interest.

It can be easily found that $d \to 0$ except at resonance, if dissipation actions are null, and $d$ is also expected to increase the higher the dissipative contribution is.

## 5. Application Examples

The damping behaviour of some multi-layer composite beam architectures with distributed visco-elastic constraints is simulated by means of the previously introduced model. Since the effect of aerohydrodynamic damping is not negligible in free and forced vibrations of large specimens [26], drag dissipative actions are linearized and modeled by means of viscous constraint actions that are distributed along the whole beam length.

Some multi-layer architectures are taken into account by considering different substrate and layer parameters (substrate geometry and material, $N$, $h_k$, $G_k$, $E_k$, $\rho_k$, $s_k$, $\tilde{\eta}_k$).

Tables 1 and 12 refer to two different beam substrate choices, while Tables 3, 4, 6, 7, 9, 10, 14, 15, 17, 18, 20, and 21 refer to different beam multi-layer solutions.

Damping estimates are evaluated with respect to the different configurations and results are shown in Tables 2, 5, 8, 11, 13, 16, 19, and 22, and Figures 3–10.

### 5.1. Configuration 1

Configuration 1 is given by a homogeneous, uniform, rectangular section single-layer beam subjected to distributed visco-elastic constraints. The mechanical parameters are given in Table 1, the first natural frequencies and damping ratios in Table 2, *d* parameter with $x_{force} = x_{response} = L$ is plotted in Figure 3.

**Table 1.** Mechanical parameters of the configuration 1 beam.

| Mechanical Parameters | Value |
|---|---|
| $L$ | 0.4 m |
| $b$ | 0.08 m |
| $h$ | 0.03 m |
| $\rho$ | $7.85 \times 10^3$ kg/m$^3$ |
| $E$ | $2.1 \times 10^{11}$ Pa |
| $G$ | $8 \times 10^{10}$ Pa |
| $k_x$ (0 < x < 0.02 m) | $2 \times 10^{17}$ N/m$^4$ |
| $k_z$ (0 < x < 0.02 m) | $1 \times 10^{16}$ N/m$^3$ |
| $c_x$ (0 < x < 0.02 m) | 400 N·s/m$^4$ |
| $c_z$ (0 < x < 0.02 m) | $8 \times 10^4$ N·s/m$^3$ |
| $c_z$ (0.02 m < x < 0.4 m) | $4 \times 10^3$ N·s/m$^3$ |

**Table 2.** First natural frequencies and damping ratios, example 5.1.

| $f$ [Hz] | $\zeta$ (%) |
|---|---|
| 178.7 | 0.77 |
| 1089 | 0.08 |
| 2928 | 0.07 |
| 3459 | 0.00 |

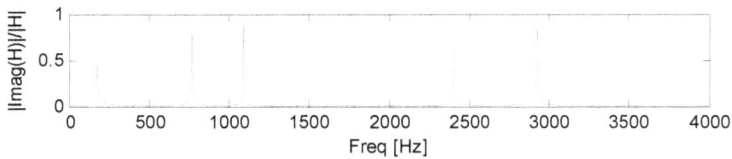

**Figure 3.** *d* damping estimator, example 5.1.

### 5.1.1. Configuration 1, Four Coating Layers, First Case

A multi-layer architecture is studied, $N = 5$, i.e., substrate + 4 coating layers. Beam length $L$, width $b$, viscoelatic stiffness constraints are reported in Table 1, and in Tables 3 and 4 the remaining mechanical parameters are listed. In Table 5, the first natural frequencies and damping ratios are given. The *d* value with $x_{force} = x_{response} = L$ is plotted in Figure 4.

**Table 3.** Layer material parameters, example 5.1.1.

| Layer | $h$ [m] | $\rho$ [kg/m$^3$] | $E$ [Pa] | $G$ [Pa] |
|---|---|---|---|---|
| $k = 1$ | $1 \times 10^{-3}$ | $8.5 \times 10^3$ | $2.1 \times 10^7$ | $9.1 \times 10^6$ |
| $k = 2$ | $1 \times 10^{-3}$ | $1 \times 10^3$ | $2 \times 10^6$ | $9 \times 10^5$ |
| $k = 3$ | | substrate: configuration 1 | | |
| $k = 4$ | $1 \times 10^{-3}$ | $1 \times 10^3$ | $2 \times 10^6$ | $9 \times 10^5$ |
| $k = 5$ | $1 \times 10^{-3}$ | $8.5 \times 10^3$ | $2.1 \times 10^7$ | $9.1 \times 10^6$ |

**Table 4.** Viscous parameters at the *k*-th interface, example 5.1.1.

| Interface | $\bar{\eta}_k$ [s] | $s_k$ [m] |
|:---:|:---:|:---:|
| $k = 1$ | $2 \times 10^{-3}$ | $1 \times 10^{-4}$ |
| $k = 2$ | $4 \times 10^{-3}$ | $2.5 \times 10^{-4}$ |
| $k = 3$ | $4 \times 10^{-3}$ | $2.5 \times 10^{-4}$ |
| $k = 4$ | $2 \times 10^{-3}$ | $1 \times 10^{-4}$ |

**Table 5.** Natural frequencies and damping ratios, example 5.1.1.

| $f$ [Hz] | $\zeta$ (%) |
|:---:|:---:|
| 172.7 | 0.74 |
| 1052 | 0.08 |
| 2823 | 0.07 |
| 3330 | 0.00 |

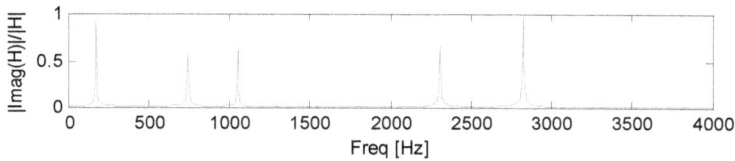

**Figure 4.** *d* estimate, example 5.1.1.

5.1.2. Configuration 1, Four Coating Layers, Second Case

A multi-layer architecture is analysed, $N = 5$, i.e., substrate + 4 coating layers. Beam length $L$, width $b$, viscoelatic stiffness constraints are reported in Table 1, and in Tables 6 and 7 the remaining mechanical parameters are listed. The first natural frequencies and damping ratios are reported in Table 8. The *d* damping estimator with $x_{force} = x_{response} = L$ is plotted in Figure 5.

**Table 6.** Layer material parameters, example 5.1.2.

| Layer | $h$ [m] | $\rho$ [kg/m$^3$] | $E$ [Pa] | $G$ [Pa] |
|:---:|:---:|:---:|:---:|:---:|
| $k = 1$ | $1 \times 10^{-3}$ | $9 \times 10^3$ | $6 \times 10^{11}$ | $2.5 \times 10^{11}$ |
| $k = 2$ | $1 \times 10^{-3}$ | $1 \times 10^3$ | $2 \times 10^6$ | $9 \times 10^5$ |
| $k = 3$ | | substrate: configuration 1 | | |
| $k = 4$ | $1 \times 10^{-3}$ | $1 \times 10^3$ | $2 \times 10^6$ | $9 \times 10^5$ |
| $k = 5$ | $1 \times 10^{-3}$ | $9 \times 10^3$ | $6 \times 10^{11}$ | $2.5 \times 10^{11}$ |

**Table 7.** Viscous parameters at the *k*-th interface, example 5.1.2.

| Interface | $\bar{\eta}_k$ [s] | $s_k$ [m] |
|:---:|:---:|:---:|
| $k = 1$ | $2 \times 10^{-3}$ | $1 \times 10^{-4}$ |
| $k = 2$ | $4 \times 10^{-3}$ | $2.5 \times 10^{-4}$ |
| $k = 3$ | $4 \times 10^{-3}$ | $2.5 \times 10^{-4}$ |
| $k = 4$ | $2 \times 10^{-3}$ | $1 \times 10^{-4}$ |

**Table 8.** Natural frequencies, damping ratios, example 5.1.2.

| $f$ [Hz] | $\zeta$ (%) |
|----------|-------------|
| 177.0 | 2.15 |
| 1083 | 1.81 |
| 2990 | 1.60 |
| 3622 | 0.00 |

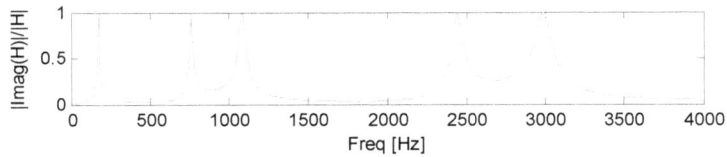

**Figure 5.** $d$ damping estimator, example 5.1.2.

### 5.1.3. Configuration 1, Eight Coating Layers

A multi-layer architecture is considered, $N = 9$, i.e., substrate + 8 coating layers. Beam length $L$, width $b$, and viscoelatic stiffness constraints are reported in Table 1, and in Tables 9 and 10 the remaining mechanical parameters are listed. In Table 11, the first natural frequencies and damping ratios are reported. The $d$ estimate with $x_{force} = x_{response} = L$ is plotted in Figure 6.

**Table 9.** Layer material parameters, example 5.1.3.

| Layer | $h$ [m] | $\rho$ [kg/m$^3$] | $E$ [Pa] | $G$ [Pa] |
|-------|---------|-------------------|----------|----------|
| $k = 1$ | $5 \times 10^{-4}$ | $9 \times 10^3$ | $6 \times 10^{11}$ | $2.5 \times 10^{11}$ |
| $k = 2$ | $5 \times 10^{-4}$ | $1 \times 10^3$ | $2 \times 10^6$ | $9 \times 10^5$ |
| $k = 3$ | $5 \times 10^{-4}$ | $9 \times 10^3$ | $6 \times 10^{11}$ | $2.5 \times 10^{11}$ |
| $k = 4$ | $5 \times 10^{-4}$ | $1 \times 10^3$ | $2 \times 10^6$ | $9 \times 10^5$ |
| $k = 5$ | | substrate: configuration 1 | | |
| $k = 6$ | $5 \times 10^{-4}$ | $1 \times 10^3$ | $2 \times 10^6$ | $9 \times 10^5$ |
| $k = 7$ | $5 \times 10^{-4}$ | $9 \times 10^3$ | $6 \times 10^{11}$ | $2.5 \times 10^{11}$ |
| $k = 8$ | $5 \times 10^{-4}$ | $1 \times 10^3$ | $2 \times 10^6$ | $9 \times 10^5$ |
| $k = 9$ | $5 \times 10^{-4}$ | $9 \times 10^3$ | $6 \times 10^{11}$ | $2.5 \times 10^{11}$ |

**Table 10.** Viscous parameters at the $k$-th interface of the multi-layer beam in example 5.1.3.

| Interface | $\bar{\eta}_k$ [s] | $s_k$ [m] |
|-----------|--------------------|-----------|
| $k = 1$ | $2 \times 10^{-3}$ | $1 \times 10^{-4}$ |
| $k = 2$ | $2 \times 10^{-3}$ | $1 \times 10^{-4}$ |
| $k = 3$ | $2 \times 10^{-3}$ | $1 \times 10^{-4}$ |
| $k = 4$ | $4 \times 10^{-3}$ | $2.5 \times 10^{-4}$ |
| $k = 5$ | $4 \times 10^{-3}$ | $2.5 \times 10^{-4}$ |
| $k = 6$ | $2 \times 10^{-3}$ | $1 \times 10^{-4}$ |
| $k = 7$ | $2 \times 10^{-3}$ | $1 \times 10^{-4}$ |
| $k = 8$ | $2 \times 10^{-3}$ | $1 \times 10^{-4}$ |

**Table 11.** Natural frequencies and damping ratios of the multi-layer beam in example 5.1.3.

| $f$ [Hz] | $\zeta$ (%) |
|---|---|
| 180.7 | 3.91 |
| 1104 | 3.62 |
| 3034 | 3.67 |
| 3622 | 0.00 |

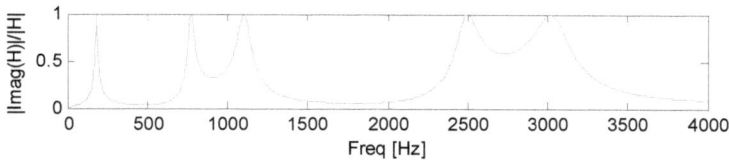

**Figure 6.** *d* estimate, example 5.1.3.

## 5.2. Configuration 2

Configuration 2 is associated with a homogeneous, uniform, rectangular section beam subjected to distributed visco-elastic constraints. The mechanical parameters are reported in Table 12, natural frequencies and damping ratios in Table 13; *d* damping estimator for $x_{force} = x_{response} = 0.3$ m is plotted in Figure 7.

**Table 12.** Mechanical parameters of the Configuration 2 beam.

| Mechanical Parameters | Value |
|---|---|
| $L$ | 0.8 m |
| $b$ | 0.05 m |
| $h$ | 0.035 m |
| $\rho$ | $2.7 \times 10$ kg/m$^3$ |
| $E$ | $7 \times 10^{10}$ Pa |
| $G$ | $2.6 \times 10^{10}$ Pa |
| $k_x$ (0.04 m < $x$ < 0.07 m) | $1 \times 10^{15}$ N/m$^4$ |
| $k_z$ (0.04 m < $x$ < 0.07 m) | $1 \times 10^{13}$ N/m$^3$ |
| $c_x$ (0.04 m < $x$ < 0.07 m) | 10 N·s/m$^4$ |
| $c_z$ (0.04 m < $x$ < 0.07 m) | $1 \times 10^4$ N·s/m$^3$ |
| $k_x$ (0.6 m < $x$ < 0.65 m) | $1 \times 10^{15}$ N/m$^4$ |
| $k_z$ (0.6 m < $x$ < 0.65 m) | $2 \times 10^{13}$ N/m$^3$ |
| $c_x$ (0.6 m < $x$ < 0.65 m) | 10 N·s/m$^4$ |
| $c_z$ (0.6 m < $x$ < 0.65 m) | $1 \times 10^4$ N·s/m$^3$ |
| $c_z$ (0.07 m < $x$ < 0.6 m, 0.65 m < $x$ < 0.8 m) | $6 \times 10^3$ N·s/m$^3$ |

**Table 13.** First natural frequencies and damping ratios, example 5.2.

| $f$ [Hz] | $\zeta$ (%) |
|---|---|
| 632.5 | 0.79 |
| 1291 | 0.38 |
| 1717 | 0.29 |
| 3191 | 0.15 |

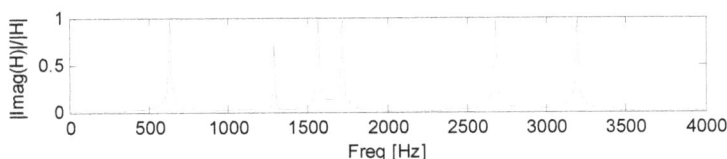

**Figure 7.** *d* estimate, example 5.2.

### 5.2.1. Configuration 2, Eight Coating Layers, First Case

A multi-layer architecture is considered, $N = 9$, i.e., substrate + 8 coating layers. Beam length $L$, width $b$, and viscoelatic stiffness constraints are reported in Table 12, and in Tables 14 and 15 the remaining mechanical parameters are listed. In Table 16, the first natural frequencies and damping ratios are given. The $d$ estimate, $x_{force} = x_{response} = 0.3$ m, is plotted in Figure 8.

**Table 14.** Layer material parameters, example 5.2.1.

| Layer | $h$ [m] | $\rho$ [kg/m$^3$] | $E$ [Pa] | $G$ [Pa] |
|---|---|---|---|---|
| $k = 1$ | $2 \times 10^{-3}$ | $3.95 \times 10^3$ | $3.6 \times 10^{11}$ | $1.4 \times 10^{11}$ |
| $k = 2$ | $2 \times 10^{-3}$ | $9.5 \times 10^2$ | $1 \times 10^7$ | $3.45 \times 10^6$ |
| $k = 3$ | | substrate: configuration 2 | | |
| $k = 4$ | $2 \times 10^{-3}$ | $9.5 \times 10^2$ | $1 \times 10^7$ | $3.45 \times 10^6$ |
| $k = 5$ | $2 \times 10^{-3}$ | $3.95 \times 10^3$ | $3.6 \times 10^{11}$ | $1.4 \times 10^{11}$ |
| $k = 6$ | $2 \times 10^{-3}$ | $9.5 \times 10^2$ | $1 \times 10^7$ | $3.45 \times 10^6$ |
| $k = 7$ | $2 \times 10^{-3}$ | $3.95 \times 10^3$ | $3.6 \times 10^{11}$ | $1.4 \times 10^{11}$ |
| $k = 8$ | $2 \times 10^{-3}$ | $9.5 \times 10^2$ | $1 \times 10^7$ | $3.45 \times 10^6$ |
| $k = 9$ | $2 \times 10^{-3}$ | $3.95 \times 10^3$ | $3.6 \times 10^{11}$ | $1.4 \times 10^{11}$ |

**Table 15.** Viscous parameters at the $k$-th interface, example 5.2.1.

| Inteface | $\bar{\eta}_k$ [s] | $s_k$ [m] |
|---|---|---|
| $k = 1$ | $4 \times 10^{-3}$ | $1 \times 10^{-4}$ |
| $k = 2$ | $3 \times 10^{-3}$ | $2 \times 10^{-4}$ |
| $k = 3$ | $3 \times 10^{-3}$ | $2 \times 10^{-4}$ |
| $k = 4$ | $4 \times 10^{-3}$ | $1 \times 10^{-4}$ |
| $k = 5$ | $4 \times 10^{-3}$ | $1 \times 10^{-4}$ |
| $k = 6$ | $4 \times 10^{-3}$ | $1 \times 10^{-4}$ |
| $k = 7$ | $4 \times 10^{-3}$ | $1 \times 10^{-4}$ |
| $k = 8$ | $4 \times 10^{-3}$ | $1 \times 10^{-4}$ |

**Table 16.** First natural frequencies and damping ratios, example 5.2.1.

| $f$ [Hz] | $\zeta$ (%) |
|---|---|
| 676.3 | 1.31 |
| 1368 | 2.13 |
| 1854 | 1.30 |
| 3647 | 1.19 |

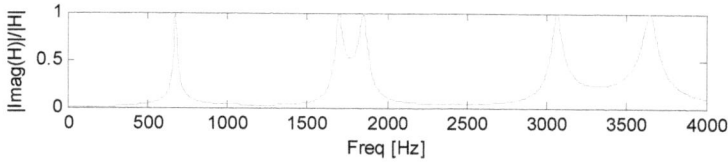

**Figure 8.** *d* damping estimator, example 5.2.1.

5.2.2. Configuration 2, Eight Coating Layers, Second Case

Tables 17 and 18 report the mechanical parameters of a $N = 9$, i.e., substrate + 8 coating layers, composite beam example, beam length $L$, width $b$, viscoelatic stiffness constraints being reported in Table 12. Table 19 lists the first natural frequencies and damping ratios. $d$ damping estimator, $x_{force} = x_{response} = 0.3$ m, is plotted in Figure 9.

**Table 17.** Layer material parameters, example 5.2.2.

| Layer | $h$ [m] | $\rho$ [kg/m³] | $E$ [Pa] | $G$ [Pa] |
|---|---|---|---|---|
| $k = 1$ | $2 \times 10^{-3}$ | $3.95 \times 10^3$ | $3.6 \times 10^{11}$ | $1.4 \times 10^{11}$ |
| $k = 2$ | $2 \times 10^{-3}$ | $1.1 \times 10^3$ | $2.9 \times 10^7$ | $1 \times 10^7$ |
| $k = 3$ | | substrate: configuration 2 | | |
| $k = 4$ | $2 \times 10^{-3}$ | $1.1 \times 10^3$ | $2.9 \times 10^7$ | $1 \times 10^7$ |
| $k = 5$ | $2 \times 10^{-3}$ | $3.95 \times 10^3$ | $3.6 \times 10^{11}$ | $1.4 \times 10^{11}$ |
| $k = 6$ | $2 \times 10^{-3}$ | $1.1 \times 10^3$ | $2.9 \times 10^7$ | $1 \times 10^7$ |
| $k = 7$ | $2 \times 10^{-3}$ | $3.95 \times 10^3$ | $3.6 \times 10^{11}$ | $1.4 \times 10^{11}$ |
| $k = 8$ | $2 \times 10^{-3}$ | $1.1 \times 10^3$ | $2.9 \times 10^7$ | $1 \times 10^7$ |
| $k = 9$ | $2 \times 10^{-3}$ | $3.95 \times 10^3$ | $3.6 \times 10^{11}$ | $1.4 \times 10^{11}$ |

**Table 18.** Viscous parameters at the $k$-th interface, example 5.2.2.

| Interface | $\bar{\eta}_k$ [s] | $s_k$ [m] |
|---|---|---|
| $k = 1$ | $4 \times 10^{-3}$ | $1 \times 10^{-4}$ |
| $k = 2$ | $3 \times 10^{-3}$ | $2 \times 10^{-4}$ |
| $k = 3$ | $3 \times 10^{-3}$ | $2 \times 10^{-4}$ |
| $k = 4$ | $4 \times 10^{-3}$ | $1 \times 10^{-4}$ |
| $k = 5$ | $4 \times 10^{-3}$ | $1 \times 10^{-4}$ |
| $k = 6$ | $4 \times 10^{-3}$ | $1 \times 10^{-4}$ |
| $k = 7$ | $4 \times 10^{-3}$ | $1 \times 10^{-4}$ |
| $k = 8$ | $4 \times 10^{-3}$ | $1 \times 10^{-4}$ |

**Table 19.** First natural frequencies and damping ratios, example 5.2.2.

| $f$ [Hz] | $\zeta$ (%) |
|---|---|
| 682.2 | 2.70 |
| 1386 | 5.38 |
| 1858 | 3.31 |
| 3642 | 3.22 |

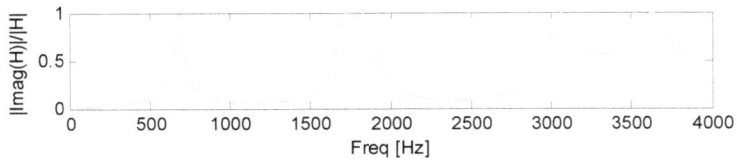

**Figure 9.** *d* estimate, example 5.2.2.

### 5.2.3. Configuration 2, Sixteen Coating Layers

Tables 20 and 21 report the mechanical parameters of a $N = 17$, i.e., substrate + 16 coating layers, composite beam example, and Table 22 the first natural frequencies and damping ratios. Beam length $L$, width $b$, viscoelatic stiffness constraints are reported in Table 12. The $d$ estimate with $x_{force} = x_{response} = 0.3$ m is plotted in Figure 10.

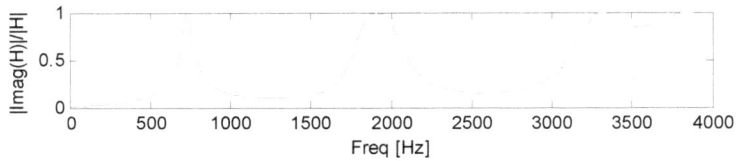

**Figure 10.** *d* plot, example 5.2.3.

**Table 20.** Layer material parameters, example 5.2.3.

| Layer | $h$ [m] | $\rho$ [kg/m$^3$] | $E$ [Pa] | $G$ [Pa] |
|---|---|---|---|---|
| $k = 1$ | $1 \times 10^{-3}$ | $3.95 \times 10^3$ | $3.6 \times 10^{11}$ | $1.4 \times 10^{11}$ |
| $k = 2$ | $1 \times 10^{-3}$ | $1.1 \times 10^3$ | $2.9 \times 10^7$ | $1 \times 10^7$ |
| $k = 3$ | $1 \times 10^{-3}$ | $3.95 \times 10^3$ | $3.6 \times 10^{11}$ | $1.4 \times 10^{11}$ |
| $k = 4$ | $1 \times 10^{-3}$ | $1.1 \times 10^3$ | $2.9 \times 10^7$ | $1 \times 10^7$ |
| $k = 5$ | | substrate: configuration 2 | | |
| $k = 6$ | $1 \times 10^{-3}$ | $1.1 \times 10^3$ | $2.9 \times 10^7$ | $1 \times 10^7$ |
| $k = 7$ | $1 \times 10^{-3}$ | $3.95 \times 10^3$ | $3.6 \times 10^{11}$ | $1.4 \times 10^{11}$ |
| $k = 8$ | $1 \times 10^{-3}$ | $1.1 \times 10^3$ | $2.9 \times 10^7$ | $1 \times 10^7$ |
| $k = 9$ | $1 \times 10^{-3}$ | $3.95 \times 10^3$ | $3.6 \times 10^{11}$ | $1.4 \times 10^{11}$ |
| $k = 10$ | $1 \times 10^{-3}$ | $1.1 \times 10^3$ | $2.9 \times 10^7$ | $1 \times 10^7$ |
| $k = 11$ | $1 \times 10^{-3}$ | $3.95 \times 10^3$ | $3.6 \times 10^{11}$ | $1.4 \times 10^{11}$ |
| $k = 12$ | $1 \times 10^{-3}$ | $1.1 \times 10^3$ | $2.9 \times 10^7$ | $1 \times 10^7$ |
| $k = 13$ | $1 \times 10^{-3}$ | $3.95 \times 10^3$ | $3.6 \times 10^{11}$ | $1.4 \times 10^{11}$ |
| $k = 14$ | $1 \times 10^{-3}$ | $1.1 \times 10^3$ | $2.9 \times 10^7$ | $1 \times 10^7$ |
| $k = 15$ | $1 \times 10^{-3}$ | $3.95 \times 10^3$ | $3.6 \times 10^{11}$ | $1.4 \times 10^{11}$ |
| $k = 16$ | $1 \times 10^{-3}$ | $1.1 \times 10^3$ | $2.9 \times 10^7$ | $1 \times 10^7$ |
| $k = 17$ | $1 \times 10^{-3}$ | $3.95 \times 10^3$ | $3.6 \times 10^{11}$ | $1.4 \times 10^{11}$ |

**Table 21.** Viscous parameters at the $k$-th interface, example 5.2.3.

| Interface | $\bar{\eta}_k$ [s] | $s_k$ [m] |
|:---:|:---:|:---:|
| $k = 1$ | $4 \times 10^{-3}$ | $1 \times 10^{-4}$ |
| $k = 2$ | $4 \times 10^{-3}$ | $1 \times 10^{-4}$ |
| $k = 3$ | $4 \times 10^{-3}$ | $1 \times 10^{-4}$ |
| $k = 4$ | $3 \times 10^{-3}$ | $2 \times 10^{-4}$ |
| $k = 5$ | $3 \times 10^{-3}$ | $2 \times 10^{-4}$ |
| $k = 6$ | $4 \times 10^{-3}$ | $1 \times 10^{-4}$ |
| $k = 7$ | $4 \times 10^{-3}$ | $1 \times 10^{-4}$ |
| $k = 8$ | $4 \times 10^{-3}$ | $1 \times 10^{-4}$ |
| $k = 9$ | $4 \times 10^{-3}$ | $1 \times 10^{-4}$ |
| $k = 10$ | $4 \times 10^{-3}$ | $1 \times 10^{-4}$ |
| $k = 11$ | $4 \times 10^{-3}$ | $1 \times 10^{-4}$ |
| $k = 12$ | $4 \times 10^{-3}$ | $1 \times 10^{-4}$ |
| $k = 13$ | $4 \times 10^{-3}$ | $1 \times 10^{-4}$ |
| $k = 14$ | $4 \times 10^{-3}$ | $1 \times 10^{-4}$ |
| $k = 15$ | $4 \times 10^{-3}$ | $1 \times 10^{-4}$ |
| $k = 16$ | $4 \times 10^{-3}$ | $1 \times 10^{-4}$ |

**Table 22.** Natural frequencies and damping ratios, example 5.2.3.

| $f$ [Hz] | $\zeta$ (%) |
|:---:|:---:|
| 730.7 | 4.46 |
| 1484 | 9.52 |
| 2000 | 5.82 |
| 3899 | 5.83 |

## 6. Discussion

The application examples presented in the previous section showed that the damping behaviour of a uniform, rectangular section beam, vibrating in flexural-axial plane conditions, may be influenced by a multi-layer coating surface treatment that is local enough to not modify the structure main geometry, strength, and stiffness.

Example 5.1.1 consists of a four coating layer surface treatment, with 2 mm upper and lower total thickness. It is shown that the damping behaviour exhibits little change with respect to the one obtained from the uncoated specimen (Example 5.1), making this specific solution not effective. As a matter of fact, the constrained layer damping (CLD) contribution [1,9], exhibited when the material shear modulus of two close layer coatings is highly different, is low in this example case. To increase the CLD contribution, in Example 5.1.2 a different, stiffer material was adopted in layers $k = 2, 4$, letting all of the remaining parameters unchanged with respect to previous Example 5.1.1. Both the first evaluated damping ratios and the damping estimator $d(\mathrm{j} \cdot \omega)$ showed an increase, making the solution related to Example 5.1.2 effective. Example 5.1.3 shows the effect of increasing the number of coating layers, by using the same layer material parameters and architecture, by doubling the number of upper and lower coating layers, and by maintaining the same coating resulting thickness. This solution is even more effective from the damping standpoint than the previously discussed architecture (Example 5.1.2).

Example 5.2 refers to a new beam substrate solution, 4 mm upper total thickness and 12 mm lower total thickness, differing from the one reported in example 5.1 by means of the material, geometry, and boundary conditions. Example 5.2.1 refers to an eight-coating layers unsymmetrical architecture, trying to maximize the CLD contribution: results are poor with respect to damping behaviour, and this solution has not been shown to be effective. Example 5.2.2 refers to the same architecture reported in Example 5.2.1, but a different, softer material was adopted in layers $k = 2, 4, 6$, and $8$, leaving all of the remaining parameters unchanged with respect to previous Example 5.2.1. An effective architecture results from the damping behaviour standpoint, since the first damping ratios and the damping functional $d(j\cdot\omega)$ increase, although the material shear modulus difference between any two alternating coating layers is lower than in Example 5.2.1. Example 5.2.3 shows that, by doubling the number of coating layers while maintaining the same architecture, same upper and lower coating thickness, the specimen damping behaviour is increased since, as expected, the contribution of the frictional actions at the layer interfaces is increased as well, so that confirming the same result obtained in Example 5.1.3.

The aim of the model-oriented approach presented in this paper is to obtain, at the design stage, an effective multi-layer coating architecture able to increase the damping behaviour of a mechanical component operating in flexural conditions without altering too much the starting geometrical, inertial, strength, and stiffness properties. Since finding a general, analytical model able to deal with any geometrical and mechanical configuration is a complex task, which requires a high number of degrees of freedom in order to be ineffective at the design stage for iteratively solving the problem at hand, a simple multi-layer beam model is proposed here.

Starting from a given uniform beam configuration, any boundary conditions and new, sub-optimal multi-layer solutions with respect to the damping behaviour may be iteratively and effectively investigated in order to be later applied to real mechanical components. Moreover, starting from the results of the present work, identification tools based on experimental dynamical measurements of simple specimens must also be developed to estimate most of the unknown layer parameters associated with the manufacturing technology, i.e., the ones related to inter-layer coupling, making it possible to extend the design optimization research stage. An experimental dynamical measurement test activity made on bi-layer specimens, mainly differing by means of the PVD, CVD, or hot melting deposition technology adopted, the substrate material, the surface finish texture, and the coating thickness, can be performed by properly designing an experimental test apparatus, following the indications reported in [27–31]. Test data can be used to evaluate the unknown interlaminar dissipation viscosity and thickness values by means of a robust, numerical identification technique, since only two unknown parameters have to be identified, following the approach developed by our research group in [11]. The multi-layer beam model may then be validated by means of new measurements related to multi-layer beam specimens and by adopting the interlaminar dissipation parameters experimentally identied in the bi-layer configuration.

Interlaminar local dissipative actions were modelled by means of a symmetrical, $C^1$ formulation reported in Equation (8), which only depended on two parameters: namely, $s_k$ and $\tilde{\eta}_k$. Since the formulation $\eta(\zeta)$ is polynomial, the evaluation of the integral in Equation (38) along $\zeta$ variable is *de facto* performed in closed form, because the integrand function $B_k\cdot\eta(\zeta)$ that appears in Equation (38) is still a polynomial function, thus making this evaluation easy and effective from a computational standpoint. It should be outlined that a non-symmetrical $\eta(\zeta)$ formulation could be easily taken into account by assuming a different upper and lower interlaminar dissipation thickness, but that an ill-defined numerical problem is expected to result when dealing with the experimental identification of the interlaminar dissipation parameters, since the number of optimization variables increases and multiple equivalent minima are also expected to result.

Results obtained by means of the multilayer beam model, consisting of the number of coating layers to be applied, coating thickness, deposition technology, and coating materials, are expected to be also effective with respect to the application to thin-walled mechanical components such as turbine blades, engine rods, and mechanism shafts among all.

## 7. Conclusions

Experimental works recently made by one of these authors and by other researchers outlined that coating layer surface treatments may increase the damping behaviour of thin-walled mechanical components vibrating in flexural conditions. Nevertheless, it was also experimentally found that most of these coating surface treatments gave a negligible or null effect with respect to the vibrational damped response in testing or operating conditions.

In order to investigate how different solutions in a free and forced flexural vibration response perform, a simple model, based on a multi-layer, zig-zag, beam assumption, which also locally models the dissipative actions at the interface between the layers, is proposed, and some virtual prototyping application examples are reported.

Numerical examples show that the beam damping behaviour can be increased by both maximizing the CLD behaviour, i.e., the relative difference of the material coating stiffness at any interface and by properly choosing the local interface dissipation parameters. A negligible effect on the damping behaviour is expected to result even if a high value of the viscous dissipation parameters at any interface between the layers is chosen, but coating material stiffness does not vary to a great extent. While the CLD behaviour is already known in principle from previous works, it appears from the reported numerical simulations that the system damping behaviour does not always generally increase by making the difference of the shear modulus of two contiguous coating layers as large as possible, but that an optimal material coating choice has to be found, so that justifying the adoption of this specific model approach. The contribution of the interaction of the layer material stiffness and the interface dissipative actions on the component damping is generally unknown at the design stage, but it can be found by means of the proposed modeling tool.

New multi-layer coating architectures may be investigated in principle by applying this modelling tool, and optimal solutions may then be applied on thin-walled mechanical components at the design stage, to reduce unwanted vibrational behaviour in high speed operating conditions.

Future research will be dedicated towards the experimental identification of the interface dissipative action parameters related to any layer deposition technology under study by testing and modelling simple dual layer beam specimen. Numerical identification techniques, based on the approach reported in [11], are currently under development by these authors. The implementation of some techniques for automatic, optimal generation of a multi-layer coating solution will also be considered and developed in future work.

**Acknowledgments:** This study was developed within the CIRI-MAM with the contribution of the Regione Emilia Romagna, progetto POR-Fesr-Tecnopoli. Support from Andrea Zucchini and Marzocchi Pompe S.p.A., Casalecchio di Reno, Italy, is also kindly acknowledged. Scientific and technical indications from Angelo Casagrande, University of Bologna, Italy, and Elena Landi, ISTEC-CNR, Italy, are acknowledged.

**Author Contributions:** Giuseppe Catania and Matteo Strozzi conceived and designed the multilayer beam model; defined and implemented the numerical model, analyzed and critically discussed the numerical results and wrote the paper.

**Conflicts of Interest:** The authors declare no conflict of interest.

## Nomenclature

| | |
|---|---|
| $L, b, h$ | beam length, depth, thickness |
| $x, z$ | longitudinal, transversal coordinate |
| $\xi, \zeta$ | dimensionless longitudinal, transversal coordinate |
| $u, w$ | dimensionless axial, transversal displacement |
| $k_x, k_z$ | longitudinal, transversal distributed elastic constraint stiffness |
| $c_x, c_z$ | longitudinal, transversal distributed viscous constraint viscosity |
| $q, F_w$ | distributed, concentrated external transversal load |
| $\rho, E, G$ | material mass density, axial modulus, shear modulus |
| $\alpha, \beta, \chi, \delta, a_k, b_k$ | kinematical variables |

| | |
|---|---|
| $\varepsilon, \gamma$ | axial, shear strain components |
| $\sigma, \tau$ | axial, shear stress components |
| $\eta_k$ | $k$-th interface interlaminar dissipation function |
| $s_k$ | $k$-th interface interlaminar dissipation thickness |
| $\boldsymbol{\phi}$ | state variable vector |
| $\Pi, U, W$ | total potential, deformation energy, external actions work |
| $\mathbf{N}$ | shape function matrix |
| $\mathbf{Y}$ | degrees of freedom vector |
| $\mathbf{K}$ | stiffness matrix |
| $\mathbf{C}$ | viscosity matrix |
| $\mathbf{M}$ | mass matrix |
| $\mathbf{F}$ | force vector |
| $\Delta\mathbf{K}$ | elastic constraint matrix |
| $\Delta\mathbf{C}$ | viscous constraint matrix |

## References

1. Rongong, J.A.; Goruppa, A.A.; Buravalla, V.R.; Tomlinson, G.R.; Jones, F.R. Plasma deposition of constrained layer damping coatings. *Proc. Inst. Mech. Eng. Part C J. Mech. Eng. Sci.* **2004**, *18*, 669–680. [CrossRef]
2. Colorado, H.A.; Velez, J.; Salva, H.R.; Ghilarducci, A.A. Damping behaviour of physical vapor-deposited TiN coatings on AISI 304 stainless steel and adhesion determinations. *Mater. Sci. Eng. A* **2006**, *442*, 514–518. [CrossRef]
3. Khor, K.A.; Chia, C.T.; Gu, Y.W.; Boey, F.Y.C. High Temperature damping Behavior of Plasma Sprayed NiCoCrAlY Coatings. *J. Therm. Spray Technol.* **2002**, *11*, 359–364. [CrossRef]
4. Kireitseu, M.; Hui, D.; Tomlinson, G.R. Advanced shock-resistant and vibration damping of nanoparticle-reinforced composite materials. *Compos. Part B Eng.* **2008**, *39*, 128–138. [CrossRef]
5. Balani, K.; Agarwal, A. Damping behavior of carbon nanotube reinforced aluminum oxide coatings by nanomechanical dynamic modulus mapping. *J. Appl. Phys.* **2008**, *104*, 063517. [CrossRef]
6. Amadori, S.; Catania, G. Experimental evaluation of the damping properties and optimal modeling of coatings made by plasma deposition techniques. In Proceedings of the 7th International Conference on Mechanics and Materials in Design, Albufeira, Portugal, 11–15 June 2017.
7. Amadori, S.; Catania, G. Damping contributions of coatings to the viscoelastic behaviour of mechanical components. In Proceedings of the International Conference Surveillance 9, Fes, Morocco, 22–24 May 2017.
8. Fu, Q.; Lundin, D.; Nicolescu, C. Anti-vibration Engineering in Internal Turning Using a Carbon Nanocomposite Damping Coating Produced by PECVD Process. *J. Mater. Eng. Perform.* **2014**, *23*, 506–517. [CrossRef]
9. Yu, L.; Ma, Y.; Zhou, C.; Xu, H. Damping efficiency of the coating structure. *Int. J. Solids Struct.* **2005**, *42*, 3045–3058. [CrossRef]
10. Zhang, X.; Wu, R.; Li, X.; Guo, Z.H. Damping behaviors of metal matrix composites with interface layer. *Sci. China Ser. E Technol. Sci.* **2001**, *44*, 640–646.
11. Amadori, S.; Catania, G. Robust identification of the mechanical properties of viscoelastic non-standard materials by means of frequency domain experimental measurements. *Compos. Struct.* **2017**, *169*, 79–89. [CrossRef]
12. Averill, R.C.; Yip, Y.C. Development of simple, robust finite elements based on refined theories for thick laminated beams. *Comput. Struct.* **1996**, *59*, 529–546. [CrossRef]
13. Cho, Y.B.; Averill, R.C. An improved theory and finite-element model for laminated composite and sandwich beams using first-order zig-zag sub-laminate approximations. *Compos. Struct.* **1997**, *37*, 281–298. [CrossRef]
14. Aitharaju, V.R.; Averill, R.C. $C^0$ zig-zag finite element for analysis of laminated composite beams. *J. Eng. Mech.* **1999**, *125*, 323–330. [CrossRef]
15. Averill, R.C. Static and dynamic response of moderately thick laminated beams with damage. *Compos. Eng.* **1994**, *4*, 381–395. [CrossRef]

16.  Di Sciuva, M.; Gherlone, M.; Librescu, L. Implications of damaged interfaces and of other non-classical effects on the load carrying capacity of multilayered composite shallow shells. *Int. J. Non-Linear Mech.* **2002**, *37*, 851–867. [CrossRef]

17.  Di Sciuva, M. Geometrically nonlinear theory of multilayered plates with interlaminar slips. *AIAA J.* **1997**, *35*, 1753–1759. [CrossRef]

18.  El-Desouky, A.R.; Attia, A.N.; Gado, M.M. Damping Behaviour of Composite Materials. In Proceedings of the IMAC XIII, SEM, 275-284, Nashville, TN, USA, 13–16 February 1995.

19.  Catania, G.; Sorrentino, S. Experimental evaluation of the damping properties of beams and thin-walled structures made of polymeric materials. In Proceedings of the 27th Conference and Exposition on Structural Dynamics 2009, IMAC XXVII, 1-17, Orlando, FL, USA, 9–12 February 2009.

20.  Mainardi, F. Fractional calculus: Some basic problems in continuum and statistical mechanics. In *Fractals and Fractional Calculus in Continuum Mechanics*; Springer: Berlin, Germany, 1997.

21.  Catania, G.; Sorrentino, S. Experimental validation of non-conventional viscoelastic models via equivalent damping estimates. In Proceedings of the ASME 2008 International Mechanical Engineering Congress and Exposition, Boston, MA, USA, 31 October–6 November 2008.

22.  Catania, G.; Sorrentino, S. Experimental identification of a fractional derivative linear model for viscoelastic materials. In Proceedings of the ASME 2005 International Design Engineering Technical Conferences and Computers and Information in Engineering Conference, Long Beach, CA, USA, 24–28 September 2005.

23.  Catania, G.; Fasana, A.; Sorrentino, S. A condensation technique for the FE dynamic analysis with fractional derivative viscoelastic models. *J. Vib. Control* **2008**, *14*, 1573–1586. [CrossRef]

24.  Paimushin, V.N.; Gazizullin, R.K. Static and Monoharmonic Acoustic Impact on a Laminated Plate. *Mech. Compos. Mater.* **2017**, *53*, 283–304. [CrossRef]

25.  Ewins, D.J. *Modal Testing: Theory, Practice, and Application*; Research Studies Press: Baldock, UK; Philadelphia, PA, USA, 2000.

26.  Egorov, A.G.; Kamalutdinov, A.M.; Nuriev, A.N.; Paimushin, V.N. Theoretical-Experimental Method for Determining the Parameters of Damping Based on the Study of Damped Flexural Vibrations of Test Specimens 2. The Aerodynamic Component of Damping. *Mech. Compos. Mater.* **2014**, *50*, 267–278. [CrossRef]

27.  Paimushin, V.N.; Firsov, V.A.; Gyunal, I.; Egorov, A.G. Theoretical-experimental method for determining the parameters of damping based on the study of damped flexural vibrations of test specimens. 1. Experimental basis. *Mech. Compos. Mater.* **2014**, *50*, 127–136. [CrossRef]

28.  Paimushin, V.N.; Firsov, V.A.; Gyunal, I.; Egorov, A.G.; Kayumov, R.A. Theoretical-Experimental Method for Determining the Parameters of Damping Based on the Study of Damped Flexural Vibrations of Test Specimens. 3. Identification of the Characteristics of Internal Damping. *Mech. Compos. Mater.* **2014**, *50*, 633–646. [CrossRef]

29.  Paimushin, V.N.; Firsov, V.A.; Gyunal, I.; Shishkin, V.M. Identification of the Elastic and Damping Characteristics of Soft Materials Based on the Analysis of Damped Flexural Vibrations of Test Specimens. *Mech. Compos. Mater.* **2016**, *52*, 435–454. [CrossRef]

30.  Paimushin, V.N.; Firsov, V.A.; Gyunal, I.; Shishkin, V.M. Identification of the elastic and damping characteristics of carbon fiber-reinforced plastic based on a study of damping flexural vibrations of test specimens. *J. Appl. Mech. Tech. Phys.* **2016**, *57*, 720–730. [CrossRef]

31.  Egorov, A.G.; Kamalutdinov, A.M.; Paimushin, V.N.; Firsov, V.A. Theoretical-experimental method of determining the drag coefficient of a harmonically oscillating thin plate. *J. Appl. Mech. Tech. Phys.* **2016**, *57*, 275–282. [CrossRef]

MDPI

St. Alban-Anlage 66

4052 Basel, Switzerland

Tel. +41 61 683 77 34

Fax +41 61 302 89 18

http://www.mdpi.com

*Coatings* Editorial Office

E-mail: coatings@mdpi.com

http://www.mdpi.com/journal/coatings